Alice's Adventures in Water-land

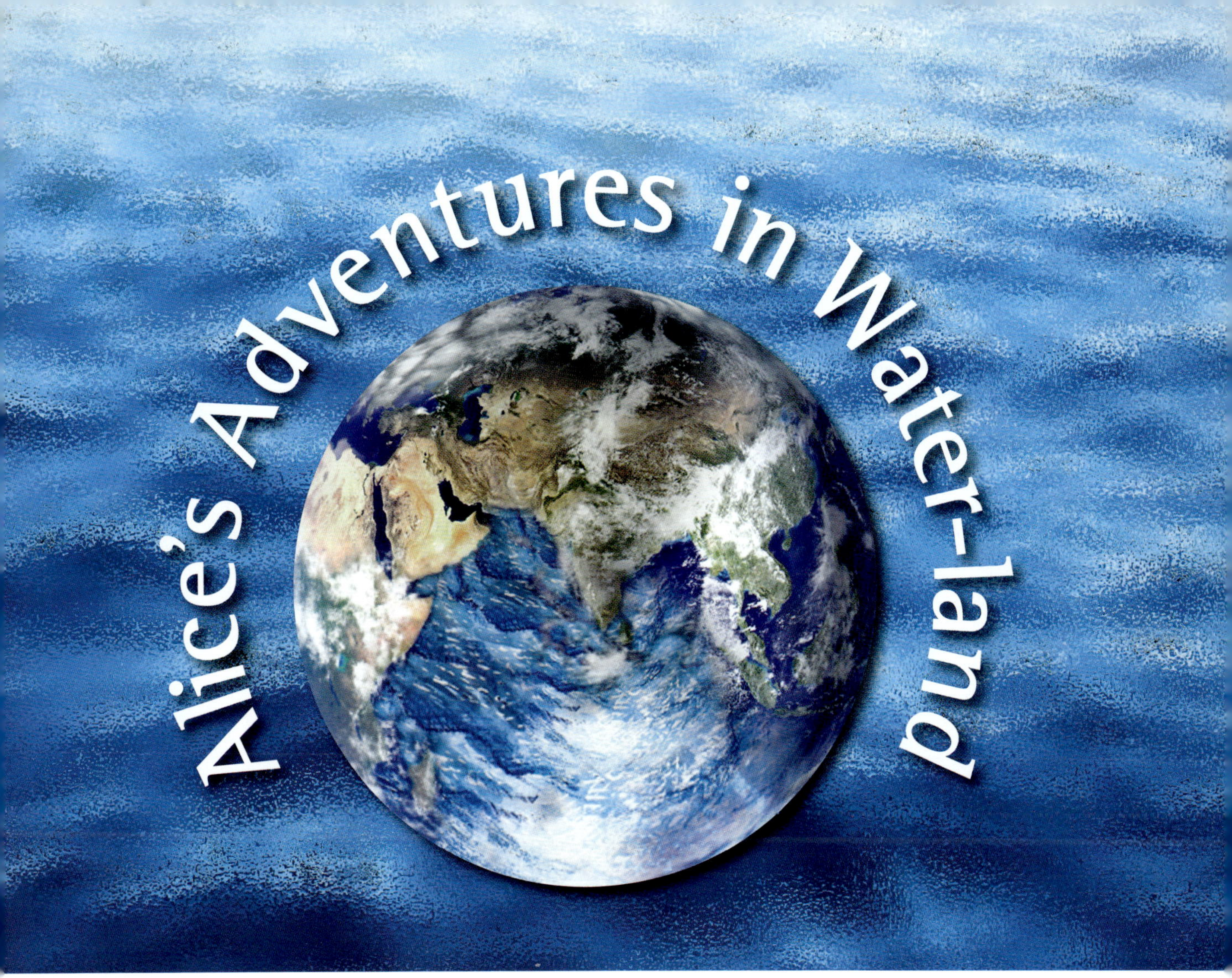

Alice's Adventures in Water-land

Arieh Ben-Naim
Roberta Ben-Naim

The Hebrew University of Jerusalem, Israel

NEW JERSEY · LONDON · SINGAPORE · BEIJING · SHANGHAI · HONG KONG · TAIPEI · CHENNAI

Published by

World Scientific Publishing Co. Pte. Ltd.

5 Toh Tuck Link, Singapore 596224

USA office: 27 Warren Street, Suite 401-402, Hackensack, NJ 07601

UK office: 57 Shelton Street, Covent Garden, London WC2H 9HE

Library of Congress Cataloging-in-Publication Data
Ben-Naim, Arieh, 1934–
  Alice's adventures in water-land / Arieh Ben-Naim, Roberta Ben-Naim.
    p. cm.
  ISBN-13: 978-981-4338-96-7 (pbk : alk. paper)
  ISBN-10: 981-4338-96-6 (pbk : alk. paper)
  1. Water--Popular works. I. Ben-Naim, Roberta. II. Title.
  GB671.B45 2011
  553.7--dc22
                              2011011045

British Library Cataloguing-in-Publication Data
A catalogue record for this book is available from the British Library.

Copyright © 2011 by World Scientific Publishing Co. Pte. Ltd.

All rights reserved. This book, or parts thereof, may not be reproduced in any form or by any means, electronic or mechanical, including photocopying, recording or any information storage and retrieval system now known or to be invented, without written permission from the Publisher.

For photocopying of material in this volume, please pay a copying fee through the Copyright Clearance Center, Inc., 222 Rosewood Drive, Danvers, MA 01923, USA. In this case permission to photocopy is not required from the publisher.

Typeset by Stallion Press
Email: enquiries@stallionpress.com

Printed in Singapore by Mainland Press Pte Ltd.

To the memory of our beloved mothers:
Rachel Ben-Naim and Loreto Batac

Contents

Preface		ix
Acknowledgments		xi
Chapter 1.	Alice's First Day in College	1
Chapter 2.	Professor Holmes Explains the Phase Map	11
Chapter 3.	The First Excursion into the Microscopic World of Water Vapor	29
Chapter 4.	The Second Visit to the Water Vapor: Experiencing the Effect of Temperature and Pressure	45
Chapter 5.	A Visit to the Solid Phase of Ice	61
Chapter 6.	Alice's Visit to the Liquid Phase	79
Chapter 7.	Why Does Ice Float on Water?	95
Chapter 8.	The Outstanding Temperature Dependence of the Volume of Water	103
Chapter 9.	The Heat Capacity of Water	115

Chapter 10.	The Latent Heat of Water and Regulation of the Body's Temperature	123
Chapter 11.	"The Electrified Water"	129
Chapter 12.	Professor Holmes' Last Lecture	139
Chapter 13.	The Last Visit to Professor Holmes' Lab	141

Preface

This book has a two-pronged purpose: first, to present the properties of the most fascinating substance on earth, and second, and most importantly, to enhance readers' awareness of the vital importance of water to life.

The book is addressed to everyone, from a child in elementary school who has just began his first steps in reading, to a retired, well-accomplished career person. There are no prerequisites for reading and understanding this book.

We hope that through our presentation of the outstanding properties of water, the reader will be encouraged to appreciate the value of, as well as to undertake a more in-depth study of, this fascinating liquid. It is also our goal to make the reader realize the vital importance of water and thus make steps, even baby steps, towards the preservation of its sources. We are, after all, beneficiaries of this wonderful gift from nature, and therefore it is our task to protect and preserve the very liquid that sustains humans, plants, animal life and anything and everything on our planet. We may start small, but if everyone makes a contribution, the results will be significant and impactful. Big achievements often begin with small efforts.

Realizing the vast spectrum of the book's target audience, we draw from the wisdom of George Gamov's writing and thus we present the material by marrying real science with science fiction (and even pure, non-scientific fiction), well-established facts with imaginative stories, and a blend of seriousness and humor. Likened to a mosaic, this intermingling of styles will, we hope, appeal to anyone who is curious and has a keen desire to explore and learn about the world in which we live.

This book was written while we spent a wonderful year at the University of Stockholm in Sweden. We are grateful to our host, Professor Aatto Laaksonen, and for a generous grant from the Wenner Gren Foundation.

For some animations relevant to this book, as well as two appendices, see www.ariehbennaim.com → books → Alice's Adventures in Waterland.

Acknowledgments

First and foremost, we are deeply indebted to Alexander Vaisman for his delightful drawings which adorn many pages of the book. Thanks are due to Guadalupe Ruiz, Kenneth Libbrecht and Bernhard Pollner for providing pictures of snow and ice. Likewise, we are grateful to Paul King, John Knight, Michael Lewis, Yizhak Marcus, Robert Mazo, Mihaly Mezei, Jorge Numata, Jerry Pollack and Samuelle Zampini for taking the time to read and comment on the manuscript.

1
Alice's First Day in College

Just like any other incoming college freshman, Alice eagerly awaited the very first day of school. She could still hear her mother's reassuring and soothing voice that morning: "Good luck, Alice." She hurriedly ran down the stairs, not wanting to be late on that very special day (Fig. 1.1).

Fig. 1.1 Alice leaves the house.

As soon as she entered the classroom and found herself a seat, Alice started to entertain pleasant thoughts of college life when a tall, lean man with a stooped posture entered the room.

"Good morning. I am Professor Martin Holmes and I am your biology professor. But before we discuss biology, we have a long way to go in order to prepare ourselves…" The professor paused momentarily. "What is the most important substance in biology?"

The class was totally silent. He waited and tried to give them a chance to answer, but sensed that no one could answer nor venture a guess.

"Water!" he impatiently blurted out.

"Now that you know the answer, we shall start with an overview of the importance of water to life. Water is not only an important and interesting liquid; it is also vital for all kinds of life."

"This semester we shall focus on the unusual properties of water, properties that have fascinated scientists for many years. In the next semester we shall discuss more specific biochemical processes where water participates."

Then he added emphatically, "What many of you probably do not realize is that the water we are enjoying right now is exactly the same water the dinosaurs must have waded through a million years ago!"

The professor's statement created quite a stir in the classroom.

"I see some shocked expressions on your faces, some of utter disbelief, but that is the reality. Water, as we use, see and enjoy it today, is the same water that has been recycled over and over again — and will be recycled in the years to come" (Fig. 1.2).

This is what the water cycle is all about.

"Where then does the cycle start? The truth is there is no starting point because the water cycle is a continuous cycle."

"For all intents and purposes, I will start with the oceans since that is where most of the water on earth exists. Without the sun, the water cycle is not possible. The sun's radiation is a major catalyst and the driving force in the heating of ocean water. The sun's radiation heats and evaporates the water on the surface of the oceans, and the vapor moves up into the air and forms clouds. Air currents also transport the vapor into the atmosphere, along with water that has been transpired from plants and evaporated from the soil. The wind moves the clouds from one region to another. When the atmospheric pressure

Fig. 1.2 The water cycle.

drops, water falls as rain, and if the temperature is low, it falls as sleet, hail or snow. In the case of snow, it can accumulate as ice caps and glaciers, which can store frozen water for thousands of years. On the other hand, snow packs often thaw when spring arrives."

"Most of the precipitation falls back where it originally came from — the ocean — and some onto the land. Due to the force of gravity, the precipitation flows over the ground as surface runoff."

"The constant movement of this water finds its way back to the ocean, and the water cycle begins again."

"Are there any of you who might have heard of the primordial soup theory?" Without waiting for an answer, he continued.

"The primordial soup is not the soup that one has for lunch or dinner, but rather a basically aqueous solution of all kinds of molecules in which the most elementary forms of life on earth have sprung. Thus, water is not only the medium in which life begun, it is also essential for sustaining life."

"Water is the lifeline of all living things — humans, animals and plants. Without water nothing can survive."

There was a certain tinge of sadness in the professor's voice as he spoke those words, and Alice thought she saw the professor's eyes well up.

The professor paused and cleared his throat.

"I opened my lecture by explaining that water we enjoy today is water that has been recycled, water that dates backs millions of years; the same water!"

Once again, there was disbelief in his students' eyes. There were hushed whispers from some who found the statement quite unbelievable.

"This is the reality," he continued. "Another sad reality is that today not everyone has access to sufficient water, which is due to several factors — natural causes or manmade factors, driven by politics."

"It is said that the amount of water used by a nation is indicative of its industrial development. Therefore, it goes without saying that the more highly industrialized a country becomes, the higher the demand for water. And the demand for water increases at a steady pace — a harbinger of dire consequences. This reality does not paint a pretty picture, does it?"

"Ensuring additional supplies of water calls for a ratcheting up of the mobilization of large amounts of resources, which more often than not, prove to be more costly. Desalination of seawater is a good example. It is now technically feasible to employ this technology, and it could very well safeguard human requirements for water forever. However, this is a costly option compared to obtaining water through other sources."

"The scarcity of water is just one issue. Due to massive industrialization, there are some areas of the world where water sources are contaminated and polluted."

"We are indeed beset with a big problem, but if we started to contribute to the conservation of water then we can help in our own simple ways. A drop in the bucket? Certainly. But if every one of us were conscious and responsible enough and did his own share then it would go a long way."

"There is a well-known idiomatic expression that I would like to share with you: 'What goes around comes around.' There's a direct connection between

Fig. 1.3 The professor discusses the 'Central Dogma.'

how we treat nature's resources today and what we will get back from nature in the future. Think about it. And now, after scaring you with a rather bleak scenario, I will proceed to the main topic for today" (Fig. 1.3).

The professor showed a slide on which was written 'The Central Dogma of Molecular Biology'. There were some hand-drawn images of proteins, DNA and some arrows indicating how molecules were transformed from one kind to another.

He began by stating that this is one of the most important discoveries of the 20th century in the field of biology. The Central Dogma explains, on the one hand, how the information contained in the DNA is replicated, and on the other, how it can be translated into a protein. He explained that the information contained in the DNA is written in a language with an alphabet of only four letters (A, C, G and T) while the information describing a protein uses an alphabet containing 20 letters (A, R, N, D, C, E, Z and so on).

The professor added that a protein that has just been systemized may be viewed as a long sequence of letters (each representing one of 20 amino acids).

6   Alice's Adventures in Water-Land

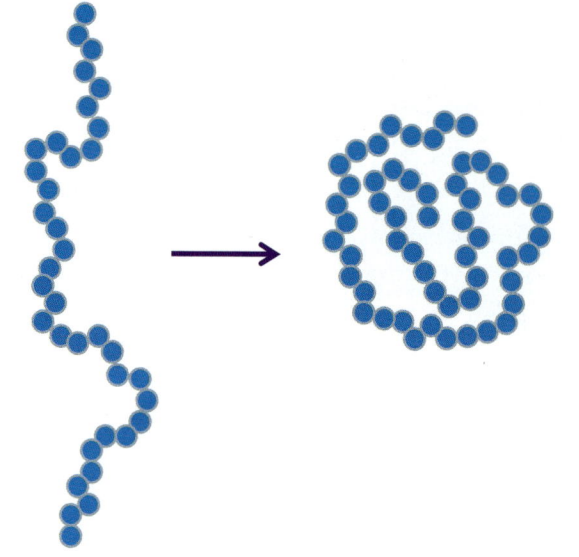

Fig. 1.4   A schematic process of protein folding.

This sequence can also be viewed as a *message*, like a sequence of letters that used to be sent by telegraph.

The protein undergoes a step-by-step synthesis in the production machinery, which is called the ribosome. Once the whole protein is produced, it is detached from the ribosome and starts wandering around, vibrating and rotating about a myriad possible axes (Fig. 1.4).

This wandering of the protein is not totally random. The molecule must reach a specific structure in a short time that is stable, relatively long-lived, and most importantly, that is the only structure in which the protein can *function* — for instance, a hemoglobin molecule carrying oxygen from the lungs to the cells, or enzymes, those micromachines that accelerate biochemical reactions.

The professor then pointed out that this process of 'protein folding' is not so well understood. A protein that has been produced from the program-carrier DNA just seems to 'know' how to fold into a specific three-dimensional structure — the only structure that enables the protein to function in the body.

But how do the protein molecules "know" the path along which to fold to the required structure? Is there an agent helping the protein? Is this knowledge contained in the sequence of amino acids, the building blocks of proteins?

"Scientists believe," said the professor, "that the water molecules that are everywhere in our body, as well as those surrounding the protein, play an important — perhaps decisive — role in the process of protein folding. To understand the role of water in protein folding in particular, and in biology in

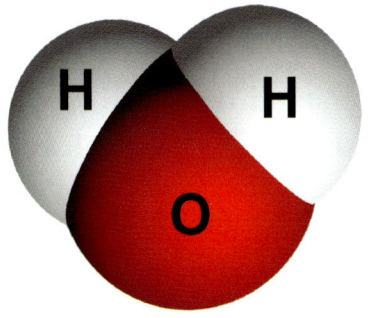

Fig. 1.5 A water molecule.

general, we must first understand the properties of water itself. Therefore, we shall dedicate most of this semester to the study of water" (Fig. 1.5).

Having laid down the course's 'grand plan,' he moved on to discuss specifics. He started by discussing the importance and the indispensability of water to life and almost instantaneously made the distinction between the properties of water viewed from the *macroscopic* and the *microscopic* points of view. The professor showed a diagram consisting of a few lines and referred to it as the 'phase diagram' (Fig. 1.6).

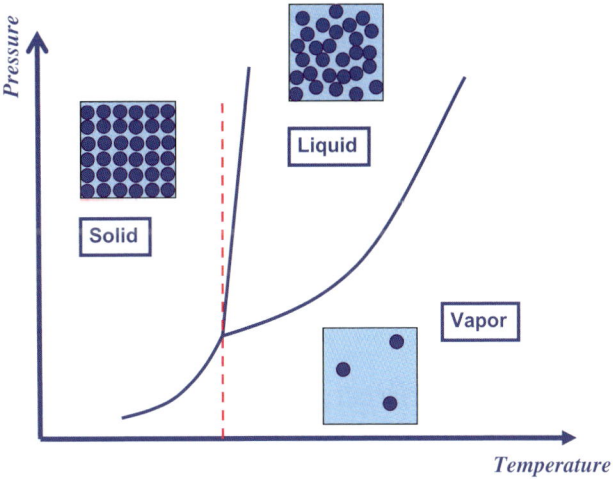

Fig. 1.6 General phase diagram.

"In this region," he said, pointing to the graph, "we have *two degrees* of freedom, whereas here, along these lines, we have only *one degree* of freedom and at this point *no degrees of freedom*."

"He is going too fast; I cannot follow him," thought Alice, her excitement turning into apprehension. Would she ever be able to understand all this?

Everything sounded Greek to her. Her stomach began churning as she tried to grasp what the professor meant.

"The microscopic point of view can help to understand the macroscopic properties, whereas the other way around is impossible," said Professor Holmes helpfully.

Lost in thought, Alice said to herself, "I can understand what the macroscopic point of view means — how we see things, hear or feel them with our senses — but what is meant by the *microscopic* point of view? Do I have to be of microscopic size to see the microscopic point of view?" (Fig. 1.7).

Fig. 1.7 Macro and micro views of ice.

In a way the professor was correct, Alice thought. In order to *see* the microscopic picture, one must be reduced to the size of an atom — which means shrinking by a factor of 1/100,000,000. Clearly, at this size the world around us would look very different from the macroscopic view.

Immersed in thought, Alice had mentally conjured up images of herself floating and then smashing against something she could not figure out, and then being blown into smithereens. The sheer thought of smashing one's self against something concrete was not pleasant and had made her wince. Being blown into tiny pieces was not pleasant either.

"But how do I do it? How will I shrink myself? Imagine Alice, be more imaginative," she said to herself. "You can do it, Alice."

But then she became unsure. "How do I know that what I imagine is really what I should be observing?"

Suddenly, she felt someone tapping her lightly on the shoulder.

"Alice, Professor Holmes is talking to you," someone whispered.

Jarred back from dreamland, Alice saw the dark, brooding eyes and the thick bushy eyebrows seemingly in a knot. The professor was staring menacingly at her.

"I will never tolerate daydreaming in my class," Professor Holmes bellowed. "I expect everyone's undivided attention. Is that understood?"

"Yes, Professor Holmes," said the class in unison.

"Class dismissed!" said the professor.

Alice was mortified. Tiny beads of sweat had trickled down her forehead. Her high expectations of her first day in college had been severely dampened by the professor's reprimand. Feeling discouraged and downtrodden, she asked herself what she was doing in the class and whether she was meant to be there at all.

The experience had sapped her of energy. Frustrated and angry at herself for not paying attention, she headed home and vowed never to be in such a predicament again.

As soon as she got home, Alice immediately sank into the comfort of her bed and only minutes after cradling her head on the pillow she had fallen into a deep slumber.

The next day was no different from the first. All of the topics that the professor had discussed were still not making any sense to Alice. She felt overwhelmed by the flood of new information. The professor talked about the different *phases* of water, showed the phase diagram once again, discussed the slopes of the curves, said something about the volume change, and… what?

"What did the professor just say?" Alice asked herself. Did I hear him say "trophy"? Trophy or was it "entropy"? Alice decided it was more like the latter: *entropy*. She had heard it before but didn't have the slightest idea what the word meant.

She looked around the classroom. Her classmates were scribbling feverishly, straining to try and catch each and every word. Alice was beginning to feel a sense of desperation, the futility of it all getting too much to bear, when the professor seemed to read her mind.

"I know these lectures are not easy to follow," he said. "If you experience any difficulties, do not hesitate to come to my office and I will do my best to help you."

Alice made a decision. "I know just what to do," she said to herself. "One of these days, I will go to the professor's office and I will ask him to explain everything to me. Then I will attend two more of his classes and if I still don't get it, I will drop the subject."

2
Professor Holmes Explains the Phase Map

The next day, Professor Holmes was subjected to a barrage of questions regarding the previous lecture. Some were discussing the topics among themselves in an animated fashion while others were just simply chatting.

Professor Holmes' incessant tapping on the desk proved to be futile. He had hoped to catch his class's attention but his tapping only got drowned in a sea of voices.

"Quiet!" he roared in a booming voice, with his signature piercing look in his eyes. That did it! The noise died down almost instantly.

"Now that I have your attention, I must tell you that our previous lecture must have left you bewildered rather than enlightened and informed. I must have gone too fast. But today, I will take you on an excursion," he said.

Clapping and stomping their feet, the class asked in unison, "Where are we going, Professor?"

"We are going on an excursion — to a place where you will encounter the three forms of water!" Professor Holmes declared, amused that he had pulled a fast one on them. Without hiding their disappointment, the students sank back in their seats.

"I promise that if you do well in my class, we will go on a field trip towards the end of the term. Now let's get back to serious business."

With that, he had managed to capture the attention of his class once again.

"Don't worry about the microscopic point of view. We shall have plenty of time to discuss that, and I am confident that by the time we get there everything will be crystal clear."

"Now, let's take a look at the *phase diagram* of water (Fig. 2.1)."

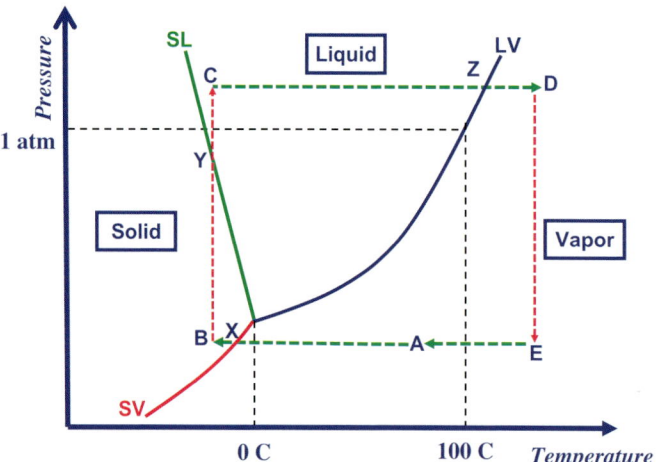

Fig. 2.1 Phase diagram of water.

At that moment, the light in the hall dimmed and up on the screen appeared a diagram with the heading *phase diagram of water*.

Professor Holmes started to explain that the word 'phase' comes from the Greek word *phasis*, meaning 'appearance' (of a star); and from *phainein*, meaning 'to show, to make appear' (as in phantasm, phantom, etc.).

"We shall use the word 'phase' here as a synonym of 'form,' 'facet' or 'appearance' of a certain substance — specifically, water. Anyone here who is not familiar with the three phases of water, raise your hand."

Seeing no raised hands, he went on, "If someone had raised their hand, I would have been both surprised and disappointed. To continue, you are all familiar with the three phases of water."

"Solid ice — these are the ice cubes that you add to your cola or the snowflakes that accumulate on the ground on a wintry day. Liquid water may be water that comes out of the tap, rain drops, and of course sea and ocean water, but most importantly, the blood that runs through our veins is mainly water. Vapor, on the other hand, is the flow of steam you see when you boil water. All substances *appear* in different forms, or *phases*. Sometimes these forms have very different properties, appearances, colors, textures, and so on. For example, carbon can be in a vapor state or a liquid state, but also, most surprisingly, two very different solid states: graphite and diamond (Fig. 2.2)."

After pausing momentarily, he added, "Diamond — this brilliant, luxurious, precious and very expensive girl's best friend — is nothing but pure carbon. For centuries, knowing full well the material and financial gains from diamonds, people have tried very hard to obtain diamonds from carbon — but

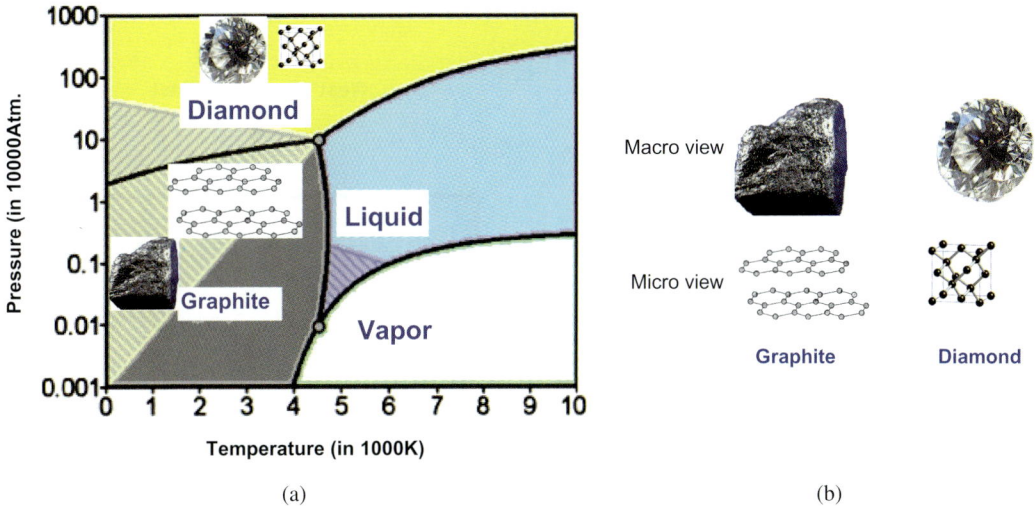

Fig. 2.2 (a) Phase diagram of carbon; (b) two forms of pure carbon.

all their efforts came to naught; they did not succeed. However, very recently some synthetic diamonds have been obtained under high temperatures and high pressures."

"Oh!" the class exclaimed, surprised and in utter disbelief.

"These two forms or phases have very different properties," continued the professor. "However, for all intents and purposes, we are not interested in carbon. Although I might add that I am personally very much interested in diamonds!"

The class roared with laughter. But, said the professor, it was time for serious business again. Alice, who had been sitting quietly in the corner, saw another side to Professor Holmes. "He is a nice man, after all, and funny too," she mused.

"Let's go back to water. What you see on the screen is the *phase diagram* of water. Actually, this is only a *small* part of the phase diagram of water — the low-pressure part. It is called a 'phase diagram' because it demarcates the regions of the various *phases* of water, i.e., the various forms or appearances of water such as vapor, liquid and solid."

Gathering her courage, Alice raised her hand to ask a question.

"Yes, young lady," said the professor.

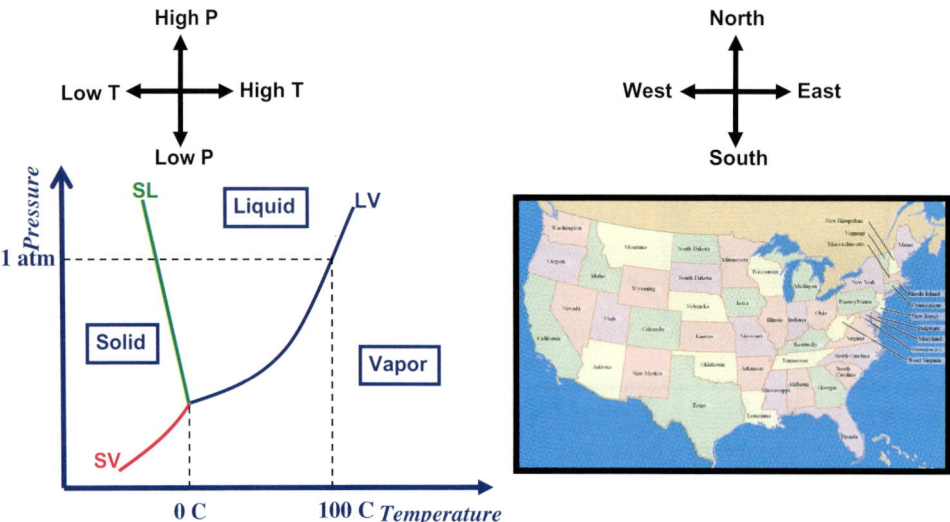

Fig. 2.3 A phase diagram and a map of the US.

"Professor Holmes, how would you visually present a phase diagram in a simpler way?" Alice asked.

"That is a very good question. In fact, I was just about to explain that," the professor replied.

"The phase diagram can be likened to a map (Fig. 2.3). Instead of going towards the east or the west, or from north to south, you go from low to high pressures or from low to high temperatures."

"You can either go in the east-west direction or in the north-south direction," he went on, pointing to a map of the United States. "We say that we have two *degrees* of freedom to walk on the map."

Someone interrupted him. "We can also go up or down in the real world, Professor."

"That is a very good comment," said the professor. "Indeed, in the real world, we can also go up and down when we descend or ascend a hill or go to the upper or the lower floors of a building. In the real world, we can move in three dimensions; we have *three degrees* of freedom. However, on this particular map, we shall restrict our movements to a two-dimensional world, left or right, or up and down. This is the same as going east-west or north-south. On this map, we do not describe the motion in the third dimension."

"However, there are phase diagrams in three dimensions — and even higher dimensional spaces. These are used to describe systems with several components, say water and ethanol. But now we are only interested in pure water, and the phase diagram can be described in a two-dimensional space."

The professor pointed at the phase diagram of water again.

"Likewise, in the phase diagram, we can imagine ourselves moving from a cold to a hot region or from low to high pressures."

"Imagine," said the professor, aiming his laser pointer at the region denoted 'vapor' in the diagram, "that you are at point A in the diagram. This point is characterized by low pressure, P, and high temperature, T" (Fig. 2.1).

"Walking within the phase diagram, up or down from point A is like treading into nothingness. The water molecules are so dispersed that it almost seems like you are walking in empty space. We shall, however, have a chance to 'see' the vapor phase from the microscopic point of view later on."

"For instance, suppose we start at point A and walk either right or left (i.e., increasing or decreasing the temperature), or up and down (increasing or decreasing the pressure). In the real world, we shall not *see* much of a change — but you will see quite a lot of changes on the microscopic level."

"You should realize that the resemblance of the phase diagram to a real map is not to be taken seriously. On a map, going to the right actually means moving east, whereas there is no real *movement* in the phase diagram. Going to the right signifies increasing the temperature of the system, not *moving* in the real world. Likewise, going up means increasing the pressure on the system."

Pointing to a point in A, in the 'vapor' region (Fig. 2.1), he continued:

"Scientists say that we have *two degrees of freedom* in this area. What this means is that we can change both the temperature and the pressure, as we wish, but we shall observe only one phase — in this case the vapor phase only."

The professor pointed to a new region denoted 'solid.'

"Similarly, in the region around this point, we can change both T and P — remember T is the short-hand notation for temperature, and P for pressure —

and as long as we do not reach one of the bordering lines, we have only one phase: *pure ice*."

Finally, he pointed to the region marked 'liquid.'

"Here, we can also change both T and P, and we will observe pure liquid water. Going to the right of this phase map means heating the water; going left means cooling the water."

"Let us go back to point A, where we have pure vapor. Suppose we keep the pressure fixed and change only the temperature. If we go to the right, like going eastward on a real map, we *increase* the temperature, which heats up the steam. In principle we can go to the right indefinitely, and of course at extremely high temperatures the water molecules can break into oxygen and hydrogen. But this is of no concern to us; these temperatures are outside the biologically relevant range."

"What happens when we cool the vapor? For as long as we do not reach the line denoted SV — short for *solid vapor* phases — there is only one vapor phase. Once we reach point X on the line SV, ice crystals start to appear. If we try to cool the system further, the temperature does not immediately change."

"What we shall observe is that more and more vapor will crystallize into ice, but the temperature will not change. We are locked in the two-phase region until *all* the vapor has been transformed into ice. Once there is no longer any vapor, the temperature will start to decrease, and from there on we move from point X into the solid phase of pure ice. Here, in contrast to what we have 'seen' in the 'vapor' region, we see only ice, solid ice."

"Once we are in the region denoted 'solid,' we recover again two degrees of freedom. Remember, this only means that we can go either left or right (changing temperature), or up and down (changing pressure), and we will still be in a one-phase region — here the solid phase of ice."

"If we continue to cool the solid, the temperature will decrease even more and we will not observe any changes in the phase of ice. But as we shall learn later on, some changes at the molecular level do occur as we lower the temperature. Eventually, we shall be approaching the absolute zero temperature, beyond which the system cannot be cooled further."

Professor Holmes noticed the students' uneasiness.

"The absolute zero temperature, written 0 K — read: zero Kelvin — is defined as $-273.15°C$ (degree celsius). All you need to know is that there exists a *lowest* temperature beyond which no further cooling is possible."

"Let us venture into another direction on the phase map. Suppose we reached point B in the solid region. We hold the temperature fixed but we increase the pressure on the system, say by adding weights on a piston. This means that we shall move upwards on the 'map.' For a while we will not observe any change; the phase will remain as solid ice. Note that when we moved from B to Y, we changed only pressure P, while keeping the temperature fixed."

"Once we reach point Y in the diagram, we shall observe changes. That is, if we try to increase the pressure further, the solid ice will start melting and we shall observe two co-existent phases, solid ice with liquid water. We say that these phases are at equilibrium, and that the point Y is on the solid-liquid equilibrium line, denoted SL. As we try to increase the pressure we shall see more and more of the ice melting and converting into liquid water. At some point, when all the ice has melted, we shall enter into the new region denoted 'liquid' in the phase diagram. This liquid is the most fascinating, as well as the most vital liquid to the whole phenomenon we call life."

A student raised his hand and asked: "How come only water sustains life? What about ice or vapor? Why can't life be sustained in these phases?"

"That is a very good question," replied the professor. "To answer this question we shall need to learn many facts about the role of water in biological systems. Also, we shall have to learn a lot about the outstanding properties of this liquid, but for now we are only interested in the *region* where liquid water exists."

"From the macroscopic point of view, that is what we actually *see* in daily life. We observe liquid water that has a light bluish color. It is so common a liquid that it is hard to believe that it has so many outstanding properties. Many scientists often refer to it as an *anomalous* liquid!"

"But let's not get carried away by the properties of water at the moment. We shall have plenty of time to be deliriously excited with this in our in-depth explorations of water. For now we shall continue to explore the different terrains of the phase diagram."

As they listened to Professor Holmes' long lecture, the students were beginning to get restless. It had started to drizzle when the professor started his lecture but now there was a heavy downpour. Some of the students had been so distracted that their minds had begun to wander. Even Professor Holmes was distracted because what had begun as rain pitter-pattering on the roof was now making it difficult for his class to hear him.

"On a rainy day like today, I think I know exactly what you are all thinking about. You would rather be curled up in bed, maybe reading a favorite book or simply gathered in the living room chatting with your families. Doesn't rain make you melancholic sometimes?"

But before any of his students could answer, Professor Holmes snapped back into lecture mode.

"So much for idle talk. Let me continue. I was talking about reaching the liquid water…"

"Oh, I see that we were moving into the liquid region. What a fascinating liquid indeed! I know you are all eager to go home but we have to finish our discussion as this will be included in your exams."

"In the liquid region, we again recover two degrees of freedom. You should remember by now that 'two degrees of freedom' is a scientist's way of saying, within this context, that we can change both pressure and the temperature independently, i.e., move right or left, up or down, and still observe only one phase."

"Suppose that from point Y we increase the pressure and reached point C in the region denoted 'liquid' in the phase diagram. We can go in two directions, which will allow us to encounter two different phenomena. We can go straight upwards and encounter some new forms of ice, but we do not need to explore those regions of extremely high-pressure forms of ice. When I say ice, I do not mean the same ice that you regularly encounter every day. I am referring to a different kind of ice; in fact, there are more than seven different crystals of ice. These ices 'live' only at extremely high pressures and are certainly not relevant to life!"

"Instead, we shall move to the right in the phase diagram and increase the temperature keeping the pressure fixed. We do that by heating the system;

heating the water causes an increase in the temperature. For a while, nothing of significance is observed; we shall be moving horizontally along the line CZ (Fig. 2.1). Once we reach point Z on the line marked LV, the liquid-vapor equilibrium line, something exciting takes place: the water starts to boil. The commonly known boiling temperature of 100°C is defined as the temperature at which the vapor-pressure of water is one atmosphere.

"But we can boil the water at a pressure other than 1 atm. If the pressure is higher than 1 atm, the water will boil at a higher temperature; if the pressure is lower than 1 atm, the water will boil at the lower temperature. Once we reach point Z, the temperature suddenly stops changing. We continue to supply heat to the system, but the temperature stubbornly refuses to change. At home, when water boils in the pot, you will notice that even when you turn the knob of the stove to the maximum setting, the water's temperature will stay constant. As we continue to supply more and more heat, water boils and transforms into the vapor phase, but the temperature stays constant. At this point, we are observing two co-existent phases — liquid and vapor. We say that the two phases co-exist at point Z on the equilibrium line, LV. Along the LV line we have only one degree of freedom. We *cannot* change both the pressure and the temperature and still maintain the two phases at equilibrium."

"What exactly does he mean by that?" Alice mused. "I wonder how the water molecules feel when they have one or two degrees of freedom?"

Professor Holmes continued.

"Further supply of heat to the system at the point Z will cause more and more liquid to evaporate. Until we reach the point when all the water has evaporated and there's nothing left of the liquid, at this point we have a pure vapor phase. Once we reach that point, further heating will no longer "boil" the liquid, simply because there is no more liquid left. Instead, the temperature starts to rise. Once again we shall be moving into the region of two degrees of freedom, and this is the region of the vapor phase denoted 'vapor' in the phase diagram. We shall be moving along the line ZD in the diagram."

At that point, Alice interrupted Professor Holmes with a question.

"Professor Holmes, how do we go back to the original starting point?" she asked.

"Good question," replied the professor. "We can easily do so by simply lowering the pressure until we get to the original pressure, point E in the phase diagram. Then we can cool the system to reach point A. In the phase diagram, we first move along the line DE, then along the line EA, to reach the original starting point A."

"We have thus completed a full cycle in the phase diagram along the lines A → B, B → C, C → D, D → E and E → A. In this excursion we encountered the three most important phases of water: the vapor, the solid and the liquid phases. The solid phase is referred to as the hexagonal, or the low-pressure, form of ice, denoted by the symbol I_h."

"Professor, why is it called hexagonal?" one of the students asked.

"We shall *see* the hexagonal structure when we discuss the crystalline structure of ice. Now, we must finish out excursion in the phase diagram. In each of the three regions we have explored there was only a single phase and within that region we were 'free' to move up or down, right or left. What is that called again?" the professor asked the class.

"Two degrees of freedom," replied the class in unison.

"We next explore the regions of one degree of freedom. These are the three lines drawn in different colors. The blue line is the liquid-vapor (LV) line, the green line is the solid-liquid line (SL), and the red line is the solid vapor line (SV)." (Fig. 2.1)

"On each of these lines we have two phases at equilibrium. As implied by the existence of the line, moving along the blue, green or red line, we are actually moving on a *one-dimensional* line. This means that we cannot independently change both the pressure and the temperature — that is, we do not have *two* degrees of freedom. The pressure and temperature change in a coordinated manner."

The professor pointed to an amplified section of the phase diagram (Fig. 2.4).

"For instance, suppose we are at point V on the liquid-vapor line, and would like to change the pressure from Pv to Pu, and to maintain the two phases at equilibrium, the temperature must change from Tv to Tu. Thus, moving along the lines of two co-existing phases, we have only *one* degree of freedom."

"You can imagine walking along a border between two countries. What will you see on each side?"

"The two countries," said the class in unison.

"Right, but here when you 'walk' on the border, you see two *phases* instead of two countries. We can change *one* of the variables — either the temperature or the pressure — and the second will be determined by the curve, as shown in the amplified section of the phase diagram." (Fig. 2.4)

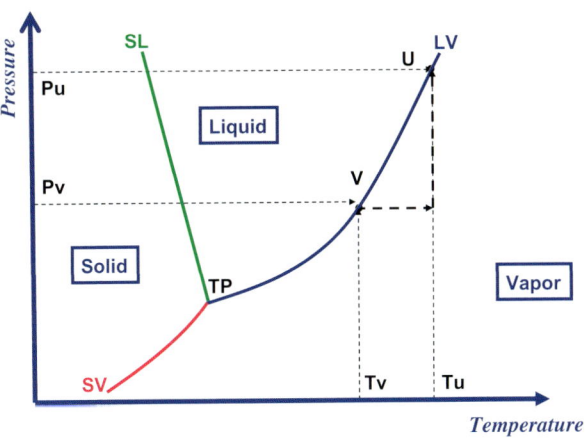

Fig. 2.4 Amplified section of the phase diagram.

"Physical chemists have found a remarkable connection between the *slopes* of these curves and the properties of the two phases at equilibrium. The theoretical dependence between the pressure and the temperature along the co-existing phases reveals a host of important and interesting properties of the phases."

"We note here that in the phase diagram of carbon dioxide (Fig. 2.5), the slopes of the three curves make an *acute* angle — i.e., an angle less than 90 degrees — with respect to the left-to-right direction of the x-axis. (An acute angle is referred to as

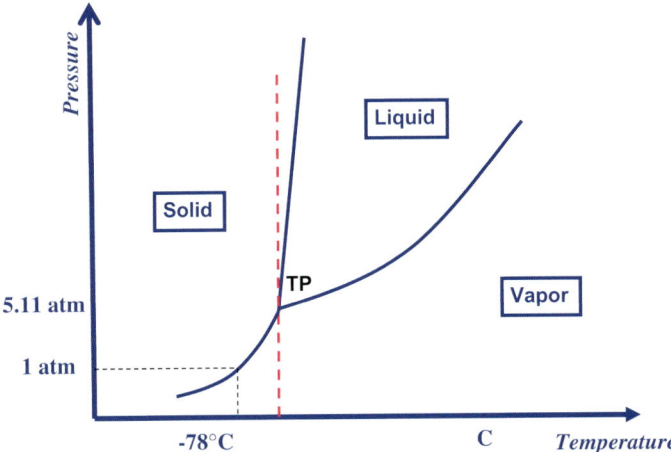

Fig. 2.5 Schematic phase diagram of carbon dioxide.

a positive slope.) This is true for both the vapor-liquid and the vapor-solid curves as indicated in the figure. However, in the phase diagram of water, the solid-liquid curve is almost a straight line and makes an obtuse angle — i.e., larger than 90 degrees — with respect to the left-to-right direction of the x-axis. (An obtuse angle is referred to as a negative slope.) This is an important observation, the implications of which will be clarified later."

"Here, I will only note that the *sign* of the slope of these curves, positive or negative, depends on the ratio of the energy required to change from one phase to another, and the change in the volume of the phase in this transition. Normally, this ratio is positive. For instance, in the transformation of a solid to a liquid phase, one needs to *invest* energy, and the change of volume is also positive (the volume of the liquid is usually larger than the volume of the solid). Water is unusual in that the melting of ice (i.e., transition from solid to liquid) requires supply of energy, but the volume of the system *shrinks* on melting."

Professor Holmes paused momentarily and then added, "I would like to suggest a small exercise. Look at each of the three curves in the phase diagram of water. Consider the transition from left to right, solid to gas, solid to liquid, and liquid to gas. Why should you expect the slopes of these curves to be positive in general?"

"At the same time, look at the phase diagram of carbon dioxide (Fig. 2.5). You will see that 1 atm pressure is below the triple point (TP). When 'dry ice' melts, it changes from solid to gas, not to liquid. Note also that 'dry ice' is not ice; it is the solid form of carbon dioxide."

"Now that we have discussed the three *lines* in the phase diagram, i.e., the one-degree-of-freedom regions, we turn to the unique point in the phase diagram, denoted the *triple point* (TP) in the phase diagram."

"As can be seen from the phase diagram, at this point the three lines meet. We have three co-existing phases: solid, liquid and vapor at equilibrium. This is a unique point in the phase diagram. This means we cannot change either the pressure or the temperature. We say that at this point we have *zero* degrees of freedom; both the pressure and the temperature are fixed. Any attempt to change either P or T would necessarily cause a disappearance of one of the phases, and we would move along the line of equilibrium between two phases."

"That is awesome," thought Alice. The idea of *one* and *two* degrees of freedom had hardly sunk in and now the professor was talking about '*zero* degrees of freedom.' She wondered how the water molecules 'felt' having no freedom. Did it mean that they cannot move at all? Were they confined, like prisoners locked up in a solitary cell?

Alice was starting to get impatient again. She had hardly processed what the professor had been explaining, and then came more and more complicated ideas. Much as she wanted to focus, her mind began to drift away from the discussion. She imagined herself in the microscopic world. 'Phases' and 'degrees of freedom' — how significant were they to water molecules? Did water molecules 'know' or 'feel' when they had one or two degrees of freedom? It all overwhelmed Alice.

She did not understand why the three lines converge at a single point. She remembered from geometry that every two straight lines meet at one point, and if the lines are curved, then the two lines can cross at more than one point. But here the professor had explained that *three curved lines* meet at a single, unique point — and no more.

After rehearsing the question in her mind a few times, she raised her hand nervously.

"Professor, how come the three curves intersect at only one point?" she asked.

"That is a very good question," replied the professor. "The answer to your question is not trivial. You are right that two curved lines can intersect at any number of points, and three lines — even straight lines — can intersect at zero, one, two or three points. But that is true if you select the lines at will.

"Here we have a different situation. The form of the curves is *determined* by the properties of the system; they are not chosen at will. It is an experimental *fact* that was known a long time ago that the three co-existing phases of a pure substance have a unique point in the phase space; this is equivalent to zero degrees of freedom. The theoretical study of this fact is encapsulated in what is now known as the *Gibbs phase rule,* since it was studied in depth by Josiah Willard Gibbs in the late 19th century. This rule summarized what we have already observed in a mathematical form. It connects the number of degrees of freedom and the number of phases. The general relationship is":

Number of degrees of freedom = 3 − number of phases

"The table shows all the possible cases for a single component," added the professor, giving the table below."

Number of phases at equilibrium	Applying the phase rule	Number of degrees of freedom
1	2 = 3 − 1	2
2	1 = 3 − 2	1
3	0 = 3 − 3	0

"You can verify that these are all possible cases. The minimum number of phases is, of course one, and that has two degrees of freedom. The maximum number of phases at equilibrium must be three because we cannot have less than zero degrees of freedom. The formula describing the Gibbs phase rule is a simple relationship between the two concepts — the number of phases at equilibrium and the number of degrees of freedom. Like many other rules or laws of physics and chemistry, this rule is a succinct way of expressing very large numbers of observable facts. I should mention here that there is an extension of this phase rule for the case of mixtures having different numbers of components, but we shall not need that generalization in our study of water."

"Before I conclude this lecture on the phase diagram, I should mention another unique and very important point in the phase diagram. This point is called the *critical point* and is denoted by the symbol CP in the diagram."

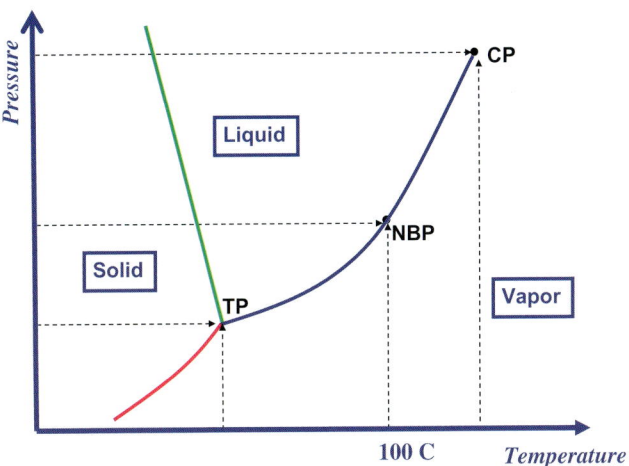

Fig. 2.6 Three unique points in the phase diagram of water.

"In Fig. 2.6, we show the three important points in the phase diagram of water: the triple point (TP), the normal boiling point (NBP), and the critical point (CP). We can reach the critical point by moving along the liquid-vapor equilibrium line — the blue line in the diagram. As we increase the temperature, the pressure also increases. Recall that along this line the vapor and the liquid co-exist."

"However, as we move up the line, the two phases have different densities. But at the critical point (CP) the two phases become identical and at that point the blue curve comes to an end. Beyond that point there exists only one phase, and no longer two phases at equilibrium."

"The critical point is a very interesting point for any liquid, not only for water. However, it is at quite high temperature and pressure — beyond the range of T and P at which life can exist. Note also that there exists no analog of the CP along the solid-liquid phase."

With this we end today's lecture. I urge all of you to read the handouts which I have given you earlier, and if there are things that you do not understand, we can discuss them when we meet again.

Alice felt a little better today although she was still not certain if she could grasp the big picture. She wanted to fully understand and not grope in the dark. So she decided to approach the professor and consult him. Immediately after the lecture she went to his office.

"I will be with you shortly as I have to talk to the convention coordinator about the arrangements for our guest speakers." He left immediately, not even allowing Alice to acknowledge what he had said.

"Professor Holmes always seems to be in hurry," Alice thought to herself. "I wonder if he will have the time and patience for me." But then he had always encouraged his students to come to him regarding things that were not clear to them. This was the only way she could thrash out all of these questions.

She looked around the professor's room and was delighted by its unexpectedly warm ambience. On the window sill were pots of beautiful orchids, the only accent of color in an otherwise sober and conventional room. Arrayed behind his desk were several pictures, some very recent, and some in sepia, mementos of his younger years.

His desk was pitifully untidy, with a mountain of books and documents piled up among the other clutter. Alice wondered how the professor could manage to work at all. A funny picture came to her of Professor Holmes buried under all his junk, and she was chuckling softly to herself when he suddenly came into the room. Alice was startled and lost some of her composure.

"Did I scare you? You seemed to be amused by something before I came in. Let me guess? It must be my cluttered desk. You must be wondering if I can work in all this chaos. YES, I can, young lady!"

They both laughed.

"So what can I do for you, young lady?" Professor Holmes asked.

She started to tell the professor how disastrous the first class went for her and that she hardly understood anything at all. She was getting nervous again and was talking too fast.

"Don't be afraid. There's no need to rush. We have enough time to talk. I realized that I went too fast with my lecture on our first day of class. Sometimes I have to be reminded," the professor said apologetically. "Otherwise I always assume that my students can follow me."

The professor's capacity to read her mind, or so she thought, never failed to impress Alice. It was as if he knew what she was about to ask before she could open her mouth. Did the professor really read her mind? Was he a psychic, perhaps?

"By coming here, you have shown your willingness to learn and that is a good sign."

"I believe I can help you in two respects. First, I will help you to *see* with your own eyes the fascinating microscopic world, which will help you understand the macroscopic world. Second, and no less important, whenever you feel you cannot grasp something, I can help you with that too, so that you feel more confident. Today and tomorrow I will be busy with some administrative tasks, but you can come to my office the day after tomorrow and I shall take you to my laboratory."

Alice left the professor's office feeling reassured and confident that with his help she would not only learn but — more importantly — understand.

She remembered the first lecture, which was such a disaster that she almost decided to drop the subject. But now after talking to Professor Holmes, her negatives feelings had dissipated. The professor was not only accommodating and helpful but he was also willing to help students like her in their quest for enlightenment and knowledge. The mere thought of being exposed to the microscopic view thrilled her no end, and while having dinner with her mother that evening, she had babbled incessantly like an excited child.

3

The First Excursion into the Microscopic World of Water Vapor

Each morning at the stroke of seven, Alice's cuckoo clock would fill her room with its melodious chime. As Alice was wont to do since she was a little girl, she would run to the window and enjoy the view. This morning ritual always invigorated her and would jump-start her day regardless of the time of year.

Looking out of her bedroom window and seeing the fog did not dampen her spirits. She had always likened the fog to a bride's tulle veil partially hiding the radiant face of a blooming bride.

"What a beautiful day!" she exclaimed.

The flowers in the planter boxes just below her window were generously adorned with dewdrops. As a small girl she had asked her mother, "Why do the leaves have tears on them? What makes them cry?" Her mother had replied, "Those are not tears, Alice. The plants are happy because you take good care of them. Those tiny, pearl-like beads of water are morning dew (Fig. 3.1)."

"Morning dew! What a lovely expression for a mere drop of water." Alice had wondered. "But where does the water come from and how come one only sees it in the morning?"

Alice was determined to unveil this mystery. Little did she know, this very topic would be discussed in her next lecture (Fig. 3.2).

After a hearty breakfast of waffles and a steaming mug of rich, creamy homemade cocoa, Alice left for school. With fond memories of her childhood filling her thoughts, Alice sauntered down the road, humming a tune and trying to remember all her questions for the professor.

Fig. 3.1 Morning due on sesuvium in Galapagos Islands.

Fig. 3.2 Dew drops on leaves.

The next lecture started with some 'excursions' in the phase diagram of water.

"Water is the liquid that we normally encounter on a daily basis," Professor Holmes had begun. "Most of the time, however, water is not pure water. For instance, seawater contains a number of solutes scattered between the water molecules. We are also familiar with the solid phase of water. We encounter solid ice as either tiny, delicate snowflakes, humongous icebergs or as the ice cubes that cool our beverages."

Professor Holmes showed a few slides of beautiful pictures of snowflakes. In some of the pictures Alice noticed the geometry of a hexagon (Fig. 3.3).

"Is that the reason ice is referred to as 'hexagonal ice'?" she wondered.

"Very dilute vapor cannot be *seen* directly. Nevertheless, we know that the air around us contains water molecules. As the temperature decreases, or as the pressure increases, the density of the water molecules increases. Eventually, when the density becomes large enough, small droplets of water molecules are formed. This is seen as fog when it occurs just above the soil or as clouds when it occurs at very high altitudes. And of course, the familiar morning dew." (Fig. 3.4)

"Dew is nothing but small drops of water that are condensed when a high density of water vapor meets a cool surface. And why only in the morning? This is the time when the temperature is the lowest, just before sunrise; when the temperature of the earth's surface starts to increase. If the temperature falls below zero degrees, then the dew turns to ice — something that we can see as frost on very cold mornings."

Fig. 3.3 Snowflakes.

Fig. 3.4 Clouds.

"Another phenomenon that you might have encountered is the coat of ice on the branches of a tree. Sometimes snow blanketing a tree turns into almost transparent ice coating the branches of the tree. This occurs when the temperature of the snow is very close to the melting point of ice. It is indeed a marvelous sight when branches are swathed in ice (Fig. 3.5)."

Hearing the professor's explanation of fog and morning dew brought a smile to Alice's lips. Since childhood, she had known fog, dew and frost, but only today did she get a really convincing explanation as to how they come about. It was truly enlightening. She was beginning to think that everything about the lecture was

 easy to understand and that she would not need to exert any special effort. But when Professor Holmes started to talk about the microscopic view, she felt her head spin. He made matters worse by saying that everything would be much clearer from the molecular point of view, or the *microscopic* view of water (Fig. 3.6).

Alice, like most of her classmates, realized it was not going to be that simple. It was clear, after talking to the other students, that no one in the class really understood what was meant by 'the microscopic view.' She thought it wise to ask the professor after the lecture. She remembered that the

Fig. 3.5 Ice on trees.

Fig. 3.6 The professor's laboratory.

professor himself had encouraged her to come to his lab and promised to help. Immediately after the lecture, she went to his office.

"How can I *see* the molecular point of view if the molecules cannot be seen, not even with the most powerful microscope?" Alice asked.

Professor Holmes smiled knowingly.

"I admire your courage and perseverance to untangle what seems to be a complicated matter," he said. "If you are really interested, I can help you understand. I have just invented two new machines, which I have never shown to anyone — although I am sure they work quite well. Now, if you are determined to understand the microscopic view of water, I can lend you my machines. They are in my laboratory. Come to the lab and see me in an hour."

Alice was glowing with excitement. "Oh my! The professor is going to lend me his inventions!" she thought to herself. "I can hardly wait! Maybe with these machines, I'll finally be able to understand what the professor has been talking about: *the microscopic view of water.*"

Professor Holmes' laboratory was housed in the basement of the science department building. The corridor leading to the laboratory was lined with plants and bathed in soft sunlight streaming through long glass windows.

Prominently displayed in the foyer was a picture of a young girl, probably only in her teens, whose arresting presence was impossible for Alice to ignore. It was as if she had met the girl with those soulful eyes before. Where? When? One day, she would ask the professor who she was.

The main laboratory was crammed with scientific instruments and rickety old desks — laden with huge piles of documents — that looked as if they would give way at any moment. On the walls, old but sturdy mahogany bookshelves cradled endless rows of books, which were presumably of great interest to the professor. The same air of homeliness and warmth that pervaded the corridor filled every corner of the aging laboratory.

Alice was looking around the laboratory when the professor's head popped out from behind a door.

"Oh, there you are, Alice! It's a nice day, isn't it?" he said cheerfully.

But even before Alice could say a word, Professor Holmes started to talk about their work.

"What I am about to show you are the fruits of my labor. I am extremely happy that I can share my inventions with someone who has the fervor and eagerness of a child," the professor declared.

Beaming with pride, but without fanfare, the professor revealed his first machine — an odd-looking telephone booth. Alice could only gasp at what she saw.

"This," said the professor, pointing to his large transparent fiberglass box, "is a shrinking machine."

"Professor Holmes must really be a genius," thought Alice, humbled that a man of such formidable talent had chosen to share his knowledge with her.

Fig. 3.7 Drops of water.

"Once you enter the box, you will experience a sensation of shrinking to the size of a molecule, and you will actually be able to watch the molecules with your own eyes," said the professor.

Alice gave the professor a disconcerted look that made him laugh.

"You have to trust me on this. You mustn't worry. Like I said, I am confident that this machine works and it's 100% safe." Then he made a quick calculation and added, "You will shrink to about 1/100,000,000 of your actual size. You will actually be able to go into the very same *phases* that I referred to in my first lecture — and you will see everything from the microscopic point of view. The first trip will take you to the gaseous phase. Are you ready?"

Alice's mind spun. Although she was ecstatic at the thought of exploring the unknown, she wasn't too happy with the idea of shrinking. What if she couldn't go back to her normal size? What if she became so tiny that bugs could swallow her whole? What if the machine malfunctioned and kept her locked inside forever?

After procrastinating for quite a while, a question suddenly popped into Alice's mind.

"What if I am there and *see* the molecular world but I still can't understand it?" she asked the professor.

"I have figured that out as well. Come with me and I will show you what I mean," he replied, leading Alice to another large structure.

It was another funny-looking phone booth.

The professor deftly removed a cloth that had been covering the booth's front panel. Underneath was a luminous blue screen. Strings of numbers danced across it. Alice could make out a faint bleep from somewhere inside the machine.

"This is the IQ machine," announced the professor. "It has two functions. First, it measures your IQ, and second, it can supply your brain with additional IQ should you need an IQ boost! All you have to do is enter the booth and it will measure your IQ. Then I will decide how much additional 'brain power' you will need to comprehend what you experience in the microscopic world, and I will charge your brain accordingly!"

"Cool!" Alice said excitedly. "It is not as scary as the shrinking machine. So will I be able to understand everything that I will experience?"

"Well," replied the professor, "perhaps not everything but quite a lot of things. You see the machine has all the knowledge that scientists have accumulated since the dawn of science. Everything that is understood today is in the machine. Of course, there are still questions or phenomena that are not well understood. For instance, scientists know enough to explain the outstanding properties of water, aqueous solutions and many other things about the function of water in biological systems. However, there are still stones left unturned. Many aspects of life are still not well understood. A classic case would be how the brain works. In fact, we are far from *understanding* how we *understand* things. In other words, our brain still does not understand how the brain works!"

"There have always been gaps in science insofar as the understanding of some phenomena is concerned, but at this point in time, we had better zero in on those aspects that are well understood rather than worrying about 'open-ended' questions."

The professor realized that this must have just added to Alice's confusion.

"Oh, what am I saying? I got carried away thinking about the problems confronting the present research efforts of scientists around the world. Please ignore what I just told you. Leave the big problems to the scientists. We have a lot of things to learn that *are* already well understood. Let's try to get to the 'heart of the water,' and you will see that you will be able to understand quite a lot."

"I have a vague idea how the IQ machine works," said Alice, "but how do you ensure that the brain power boost that you give me is enough to allow me to get the picture when we are sent on our mission?"

"I have to calibrate the machine on a monthly basis. On the first day of the month, I measure the IQ of all the students in the class, take the average and this will be the IQ 100. Every reading above that means a higher than average IQ. On the other hand, every reading below that means, well, not enough IQ to understand the mission," Professor Holmes replied with a chuckle. "Now, are you ready?"

"Ready when you are, but what am I supposed to do?"

"There is not much you have to do today, and as I see it you do not need the IQ machine for now. At a later time, however, when something seems to be beyond your reach then you can use the IQ machine. By the way, the password that will transport you back is IWANTTOGOBACK."

The professor explained to Alice that any time she feels she wants to go back to the real world, all she has to do is to say the password and she will return to her original size.

"Follow me," the professor commanded.

Putting on goggles as the professor had instructed her, Alice stepped into the shrinking machine and sat down on the small stool inside. When Professor Holmes shut the door, Alice was tempted to back out. Imagining the worst-case scenarios sent the butterflies in her stomach fluttering wildly. Her mild palpitations turned into a deafening, thumping heartbeat. It felt as if her chest was going to explode.

"Be calm, Alice. Everything will be alright," she thought.

Then, in an instant, everything in Alice's world changed. She began seeing what the professor had described. Strange, two-legged creatures were suddenly flying all around her. They must be water molecules! Some were slow and some were fast. They rotated and spun around in all directions as they flew. It was as if the molecules were doing some kind of exercise! She followed the rhythm of their movements and started to count — one, two, three! — as the creatures stretched and contracted their arms, opening and closing the angle in between (Fig. 3.8).

To get a better picture of what they were doing, Alice climbed on top of one of the molecules. The first thing that struck her was that although the molecule

Fig. 3.8 Alice in the vapor phase.

was flying fast, she did not have the 'windblown' experience she normally had when she traveled in a car at high speed. It was strange that in spite of the high speed, there was no blasting of air against her face. She made a note of that as well as the stillness of it all. Although there was motion at high speed, there was no sound at all.

Alice was awestruck. The experience of flying at speed in total tranquility was overwhelming. Every so often another molecule came close to the molecule on which she sat, and collided with it, abruptly diverting her course. A few times the collisions were so violent that they almost threw her off the molecule — and the spinning of the molecules was making her feel quite dizzy. The molecules spun around three axes, and she had to anchor herself to the molecules' arms to stop herself from flying off!

Holding the molecule's 'arms,' Alice observed how they exercised. One exercise involved opening and closing the angle between the two arms, a relatively slow motion compared to the faster motion of stretching and contracting. Sometimes the motion was symmetrical, with the two arms stretching and contracting in concert with each other. At other times they would move asymmetrically — one arm stretching; the other contracting.

"These must be the fundamental vibrational modes," Alice recalled from one of Professor Holmes' earlier lectures.

Alice noticed that there were just these three kinds of motions, along with the occasional collision. Between collisions, the movement was in a straight line, and each collision caused an abrupt change of direction. The vibrational 'exercises' of the molecules were regular and not really affected by the flight speed or by the dizzying rotations.

Alice thought about the 'two degrees of freedom.' Where were they? She would have to clarify this matter with the professor when she got back to her world… Suddenly, Alice realized something awful. In all her excitement, she couldn't remember the password! Professor Holmes' enthusiasm had got her so fired up that she had completely ignored the part about getting back again!

"How could I forget the password?! I might never be able to get out of here!" she cried.

Over and over, Alice cursed herself for not paying attention. Her heart pounded and rivulets of tears streamed down her face. In desperation, she screamed, "I want to go back!" — praying that somehow the professor would hear her and come to rescue her.

Suddenly, the molecules around her started getting smaller and smaller. In no time at all, they had completely disappeared. Alice realized that her body was back to its original size and she was standing in the pitch black of the booth. She took off her goggles and heaved a huge sigh of relief when the door opened revealing Professor Holmes' friendly face.

"Thanks for coming to my rescue," said Alice. "I was so worried that you wouldn't hear me!"

"Don't tell me, you forgot your password?" laughed the professor, looking at Alice's ghastly appearance. "I thought that you might forget! That was precisely one of my considerations when I made the machine. I did not hear you, but I devised the password in such a way that merely calling for help would be enough to let one out."

Alice was astonished. "But if you didn't hear me screaming, how did you know I needed help?"

"Of course I didn't hear you. I couldn't have heard you while you were inside the machine. You should know that the voice, or any sound for that matter, travels through the air, and since there was no air between the molecules you were visiting, there was no medium to carry the sound," said Professor Holmes, amused by his made-up explanation.

The professor went on to explain that he designed the machine taking into account the person forgetting the password and screaming for help and those words conforming to a valid password.

"You were part of the air so you could not send sound waves to communicate between molecules." Professor Holmes smiled impishly and added, "You will understand better when you will know more about how this machine works."

Alice still could not understand and she was a bit disappointed in herself. But the idea of sound traveling through the air interested her. She had already understood two of the puzzling observations that she had wanted to consult the professor about. The first was the absolute stillness that enveloped the molecules. The second was the lack of 'air resistance' she had experienced — a direct contrast to the sensation of 'wind' one feels when one travels fast in the real world. It had become clear to her that she had visited the atoms themselves, and that the spaces in between them were completely empty. There was no 'air' around her; she was *part* of the air. And since sound travels *through* the air, there was no way to transmit sounds into the emptiness, between the molecules.

"How could she breathe in that kind of environment?" Alice wondered. But first she needed to clear up the matter of the 'two degrees of freedom.'

"While in the gaseous phase of water," Alice said. "I recalled your lecture about moving left or right, up or down in the phase diagram. You told us that these motions correspond to changes in the pressure, P, and temperature, T. But although I was flying fast and probably covering quite a distance, I did not *feel* any change in pressure or temperature."

"You are perfectly right in your observation, and you deserve an answer," the professor replied. "It is quite obvious that you are still confused about the macroscopic and the microscopic worlds. These are two very different 'worlds.' Although you have visited the microscopic world, you have not experienced enough, and the task at hand is to learn and think in different terms when you see the microscopic view."

"For instance, the concepts of pressure and temperature have very different meanings, or perhaps the right thing to say is that they are perceived differently in the two worlds. When you were there, you were actually at *one point* in the phase diagram. True, you were traveling at high speed and covered a vast distance in a short time, but that distance is different from the *distance* we cover in the phase map. In reality, you actually stayed at one fixed point."

"I suggest that you go home now and rest, and try to digest what you have learned so far. Next time we will look at the excursions on the phase map from the molecular viewpoint."

On her way home, a light breeze gently stroked Alice's face, and she wondered whether she was feeling water, oxygen or nitrogen molecules. The sensation of the wind was so different from what she had experienced being part of the wind itself.

4

The Second Visit to the Water Vapor: Experiencing the Effect of Temperature and Pressure

The melodious chime of the cuckoo clock roused Alice from her sleep. Looking around her room, which was bathed in the morning sun, she exclaimed, "What a beautiful day!"

Alice was still overwhelmed by the experience of the previous day. It seemed she now knew some secrets no one else was aware of — except, of course, the professor, the inventor of the shrinking machine.

She was eager to learn more and to experience first hand what the professor meant by the two degrees of freedom. She understood that the phase map was merely a metaphor for the real phase diagram. She also understood what it meant to increase or decrease the pressure and the temperature. In fact, not only had she understood, but she had actually experienced it with her senses. But how were these concepts of pressure and temperature *seen*, or perceived, in the microscopic world?

Alice was unusually quiet at breakfast, and it had not escaped her mother's attention. The girl seemed to be lost in thought over some unknown problem.

"What's bothering you, sweetie? You look as if you're carrying the weight of the world on your shoulders," said Alice's mother, finishing her sentence with a chuckle.

"Oh Mom, don't be silly. I was just thinking about our lesson the other day. The professor explained things so clearly that I can't think about anything else. Science classes are rarely so interesting and exciting!"

Fig. 4.1 The gadget.

She did not tell her mother about the shrinking machine as she wasn't sure how her mother would react. Besides, she had no idea how it worked so she was in no position to explain anything — especially to her mother.

"I'd better get going, Mom. Professor Holmes promised to help me out with some experiments." Alice picked up her plate and mug and put them in the dishwasher.

On the way to the laboratory, Alice rehearsed her questions for the professor. Learning something new each day was fueling her enthusiasm to find out more. As she entered the foyer, the mysterious girl in the portrait welcomed her again. But before she could ask a single question she had written down her on her notepad, the professor appeared and handed her a gadget. The device had two knobs marked with the letters T and P (Fig. 4.1).

According to the professor, with these two knobs Alice would be able to navigate through the TP map. Turning the T knob to the right was equivalent to increasing the temperature while turning the P knob to the right increased the pressure.

"These two dials will allow you to manipulate the two degrees of freedom," explained the professor. "In the real macroscopic laboratory we can take a sample of gas — water or any other substance — and we can control the pressure on a piston by adding different weights (Fig. 4.2), or the temperature by heating the system (Fig. 4.3). As you can very well remember, each point on the phase map corresponds to a point in the TP diagram — one specific temperature, T, and one specific pressure, P. Let me explain the correspondence between the macroscopic experiment we carry out in the laboratory and the *virtual motion* in the phase diagram."

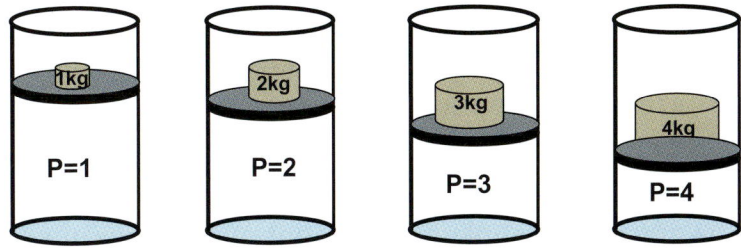

Fig. 4.2 Effect of pressure on the volume of a gas.

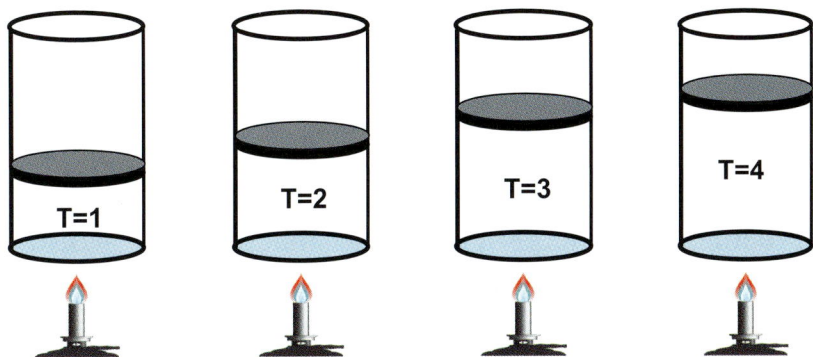

Fig. 4.3 Effect of Temperature on the volume of the gas.

The professor showed her a cylinder sealed with a piston. On the piston were a few weights, and under the cylinder, a burner.

"By adding or removing weights on the piston," the professor went on, "we can increase or decrease the pressure in the gas contained in the cylinder. This corresponds to the virtual motion, up or down (or north or south) in the phase map. On the other hand, we can either heat or cool the gas — i.e., the temperature will increase or decrease. This corresponds to the virtual motion east or west in the phase map. All these correspond to *macroscopic* experiments. Your task today will be to experience the *microscopic* view of these experiments."

Alice was very clear about the macroscopic experiment, but she could not imagine what the corresponding microscopic experiment would look like — or feel like. She wondered how she would *feel* the high pressures when she turned the P knob, and whether she would feel cold or warm when she turned the T knob. Would turning either one to the extreme left or extreme right make her feel very cold or very warm? She found the idea quite frightening.

The professor seemed to have read Alice's mind. Even before she could utter a word, he started to explain that what she *experienced* in the microscopic world would be very different from how she experienced pressure and temperature in real life. In the macroscopic world, you *feel* the temperature and the pressure through your skin, since there are nerve cells under the skin that respond to changes in pressure or temperature. These send messages to the brain, where they are processed to produce either the sense of pressure or temperature. You can also *measure* the temperature and the pressure using the appropriate instruments — the thermometer and the barometer.

"Human beings have experienced the macroscopic world since time immemorial. They have sensed the temperature and the pressure of the air. But experiencing temperature and pressure is one thing; understanding them is quite another. What you are going to explore today is quite different from the macroscopic experience you encounter in the laboratory. The microscopic view of vapor is based on an *understanding* of the pressure and temperature of the gas at a molecular level."

"To experience the microscopic view, you have to use the shrinking machine, observe and take notes of what you see. However, to be able to absorb and fully grasp what you experience, you will need some additional 'brain power!' I say this not because I have doubts about your intelligence, but because I know that for almost 2000 years these phenomena were not well understood at a microscopic level. Even after two giants of physicists, James Clerk Maxwell and Ludwig Boltzmann, explained these phenomena it was not easy to convince even their fellow scientists of the late 19th century."

"Would you like to experience the microscopic view first and leave the understanding for later? Perhaps, it would be a good idea if you observe first and then we can discuss it later. That way you will not be overwhelmed. You

should know, however, that no scientist has ever experienced the microscopic view. My machine is only a simulated experiment based on what scientists have learned," the professor added.

Alice decided to forgo the use of the IQ machine. She could always use it some other time.

Alice was very eager to experience the microscopic world — so eager, in fact, that she was getting quite impatient. She was confident that her brain — without being 'charged' — would be able to absorb and process everything. Gone was her feeling of creepiness as she stepped inside the shrinking machine. She knew what to expect this time and no longer found the machine daunting.

All of a sudden, she was back in the microscopic world. Everything was exactly the same as in the previous day's encounter — water molecules were flying in straight lines, vibrating rotating, rotating and occasionally colliding with one another — but today's scene seemed so real to Alice. A day earlier, the molecules were like images painted on canvas, but this time those very same images seemed to have come to life, vibrant and pulsating.

"This is getting so exciting," said Alice to herself. "I am really getting to love this!"

She felt as if she could actually reach out to those images, feel and touch them. Deeply absorbed by what she had witnessed, she almost forgot why she was there in the first place until she remembered the TP gadget. She decided to use it and find out for herself how it worked.

Alice started with the P knob, turning it slowly counterclockwise to a lower setting, which meant lower pressure. As she turned the dial, there was no discernible change at first although she noticed that there seemed to be fewer water molecules and the collisions between the water molecules had become increasingly infrequent. She turned the dial all the way down to almost zero. Most of the water molecules disappeared. It was almost total emptiness — a vacuum! Only once in a while would a molecule cross the chamber where Alice sat.

When Alice finally reached zero on the P dial, all the water molecules had vanished, and she felt completely alone. Was this the way it felt at low pressures? She would ask the professor about it, or perhaps she would tell him how it feels. She wondered where all the molecules had gone.

Alice then turned the P knob clockwise and the water molecules started to appear. As she continued to turn the dial towards higher pressures, she noticed that more and more water molecules appeared, and the collisions occurred more frequently. At times, two molecules collided, and sometimes even three or more molecules. She did not *feel* any *pressure* change although the scale showed first 2 atm, then 3 atm, and before long, 10 atm. She had also expected her ears to pop with the pressure, but felt nothing. Some of her friends who went diving had told her about the high pressure at depth, the dangers, and the possibility of feeling pain when they returned to lower pressures at the surface.

Alice was getting excited at the prospect of learning more. She slowly increased the pressure, and the density increased. As she watched the molecules, she noticed that a few of them joined together as if they were holding hands — which reminded her of skydivers holding hands as they take the plunge. First pairs, then triplets, then more and more water molecules seemed to hold hands and remained bound to one another — instead of colliding and dispersing again.

As she carried on turning the dial, she saw clusters of water molecules going downwards in free fall. She followed one of the clusters as it reached the bottom, where it was filled with densely packed water molecules. She thought that it could have been point Z on the phase diagram that she had heard about in one of the lectures. She figured out that she was approaching that point from the gaseous phase, i.e., moving upward in the phase diagram.

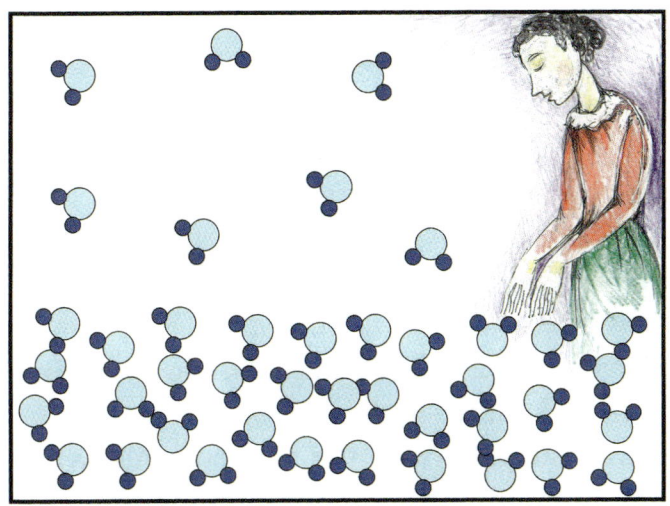

Fig. 4.4 Alice observing the gas and liquid phases.

Alice then tried to further increase the pressure, but the density of the water in the vapor phase remained unaltered. Instead, more and more water molecules 'condensed.' Bonding with one another, they sank to the bottom where a liquid phase had formed. For a while, Alice thought that she had reached the liquid-vapor coexistence line that the professor had talked about. But now she was more certain that she had hit point Z in the phase diagram.

She did not want to continue to increase the pressure, but she was curious to see how the T dial would change things. She turned the P dial back to its original position, figuring that she was at one of the points near A in the phase diagram. Then she tried to turn the T dial clockwise to increase the temperature, but much to her surprise, she did not feel any warmer. She turned the dial up a notch, but still no feeling of warmth. She would have to tell the professor that the T knob wasn't working.

Suddenly, she realized that after increasing the temperature the molecules had begun to fly faster. She gingerly extended her arm and touched one of the molecules. It didn't *feel* any hotter than before. What was going on?

Alice decided to explore the low temperature region. As she turned the knob counterclockwise, everything seemed to start moving in slow motion. The molecules began flying at a slower pace, and they were even spinning slower. Yet, strangely, Alice did not feel cold at all.

"Did the professor give me the wrong gadget?" Alice wondered.

She distinctly remembered her conversation with the professor — he had said that the T dial controlled the *temperature*. But now it looked as if the knob

was controlling the *speed* of the molecules instead. She did not feel any change in temperature. Perhaps the professor meant that this button controlled the *temperament* of the molecules, and not the temperature! Alice made a note. She would be sure to clarify with the professor.

Alice continued to turn the T dial counterclockwise and the molecules slowed down more and more. She wondered what slowing down had to do with the temperature. As she was trying to figure out the connection, she noticed that at the bottom of the chamber was a thin layer of water molecules that were arranged in an orderly fashion, filling the surface with hexagons. Each molecule seemed to be attached to four other molecules. She turned the T dial even further, hoping to see something different, but nothing happened to the gas molecules in the gas phase. They continued to move and rotate at the same speed.

"That's rather strange," thought Alice to herself. "Earlier, when I turned the T knob counterclockwise, there was an overall slowing down of the motion of the molecules. But even that slowing down process has now ground to a halt." She continued to turn the knob counterclockwise but to no avail. Nothing tangible seemed to materialize.

Fig. 4.5 Alice observing the gas and solid.

Beginning to get frustrated, Alice looked down and was astonished to see that the layer of hexagons at the bottom had sprung up. An increasing number of molecules from the gas phase were diving towards the bottom and seemed to be searching for the right spot to join the orderly arrangement of hexagons. The more she turned the T dial, the more molecules joined the beautifully arranged hexagonal tiles. She also noticed that despite turning the T dial, the motion of the molecules did not change. It reminded her of what the professor had said in his lecture when he described point X in the phase diagram.

"Was there some mistake? Was this T dial for *tiling*?" Alice thought to herself. "I must clarify this with the professor."

Tinkering with the T dial, she noticed that turning the button clockwise caused the molecules to take off from the surface as they regained their free motions of translation and rotation. On the other hand, as she turned the dial the other way, the molecules landed on the surface and formed a layer of regular structures, one layer on top of another.

The observations baffled Alice, and she had amassed quite a collection of unanswered questions. What did all this have to do with the mere turning of the T dial? Why did they take-off from the surface as she turned the dial to the right? There did not seem to be any contraption to catapult the molecules. Was she in some way providing the energy to kick the molecules into the gaseous phase?

Alice was eager to learn more, yet impatient to fully grasp what it was that she had actually observed. More stones unturned, more puzzles to solve. What was the T dial really doing? She had to solve these mysteries first before discovering more. It was time to say the password.

"I WANT TO GO BACK!" yelled Alice, this time without trepidation. In an instant, the molecules were once again reduced to tiny specks until they completely disappeared. There was an eerie silence — so quiet it was deafening — and in the pitch darkness, Alice suddenly felt a little scared. Then a narrow beam of light emerged from the gloom and before she could even adjust her eyes, there was a sudden burst of blinding light. And who else did Alice see but the smiling, welcoming face of the professor.

"You have lots of questions to ask, am I right?" asked Professor Holmes, smiling.

"Yes, indeed, Professor. I saw so many things that I did not understand. And I also think that you gave me the wrong gadget! The T knob didn't work the way you told me it should, although I followed your instructions to the letter. Or could it be possible that you gave me the wrong instructions? I did not feel cold nor did I feel warm at all? What happened? Did I do something wrong?" Alice asked.

In his mischievous tone, the Professor said, "That's normal. You really could not understand everything and there's nothing wrong with that. What you just observed is something that no human being has ever observed. You were very lucky to be able to use the shrinking machine and see the world for yourself — first hand — from a microscopic point of view. For many years scientists had only imagined how the world would look from the molecular level, if a molecular level existed at all."

"Let me offer you some options. I can either give you a formal lecture and I will try to answer your questions, or I can provide you some facts and findings of scientists from the late 19th century. You will enter the IQ machine to boost your brain power so that you will be able to understand all of the phenomena that you have experienced. But what you saw today has given you an edge. What shall it be then?" the Professor asked.

Alice was filled with uncertainty. Making choices did not come easy. She also had qualms about the professor's IQ machine as it might just reveal how low her IQ was! In spite of her misgivings she decided to give the machine a try. Regardless of her IQ, finding out what had happened and understanding was her priority!

"All right Alice, I will give you some background information and with the help of my IQ machine you will be able to understand what's going on," the Professor said reassuringly.

Inside the IQ machine, Alice braced herself for the sudden flood of knowledge into her brain. She needed to make a conscious effort to be alert and acutely aware so she could really feel what the machine was going to do to her.

But she felt nothing at all.

She would have to ask the Professor later what this strange machine was doing. As she stepped out of the machine — feeling exactly the same as when she had entered it — she heard the Professor's words:

"For more than two thousand years, people contemplated the nature of matter. Suppose you take a piece of iron and cut it into two. You get two halves of the same substance. You repeat the procedure, each time cutting the material into two. Will this process ever come to an end?"

"Well, you can cut the substance into halves ten times or a hundred times, or a million times, but at some point you can no longer continue to handle those tiny pieces. There is no tool that can handle minute pieces of iron. Further division is only a 'thought experiment' and the best we can do is to *imagine* what will happen if we continue the process indefinitely."

"But this is purely down to our imagination. No one can claim with certainty how many times we can repeat the cutting process, and whether or not this process will come to an end. Scientists have speculated that after some time we will reach the point where the pieces can no longer be cut. They called these pieces 'atoms,' from the Greek *a'tomos*, which means uncuttable.

The Greek philosopher Democritus coined the term some 2500 years ago, and the existence of such atoms remained a mere hypothesis for a long, long time. No one could prove that they existed. Two thousand years after the Greeks first speculated about atoms, scientists still did not have concrete proof of their existence. However, some indications of their existence had been slowly accumulating — not directly by observation, but rather indirectly. By postulating the existence of atoms and molecules, scientists were able to explain the properties of gases, including the meaning of pressure and temperature and the relationship between volume, pressure and temperature known as the 'equation of state of ideal gas.'"

"Today, we know that atoms are the smallest units of pure elements such as oxygen, hydrogen or iron. Atoms combine to form molecules, which are the smallest units of a substance. For instance, a water molecule consists of two atoms of hydrogen (denoted H) and one atom of oxygen (denoted O). A water molecule (denoted H_2O) is thus the smallest unit of the substance we call water."

The Professor showed Alice a cylinder mounted on top of a moving piston, the system wholly submerged in a water bath.

"In the cylinder, there is a gas of water molecules," he explained. "By pressing on the piston — or by adding different weights — we can change the *pressure* on the gas, and by heating or cooling the water bath, which we call a thermostat, we can change the temperature of the gas within the cylinder."

"One of the most remarkable results of studies on this simple system was the realization that the pressure exerted on the piston from the *exterior* is balanced by a pressure exerted by the *molecules* on the piston from the *interior*. Bernoulli was the first to explain that pressure is produced simply by the frequent bombardments of a huge number of molecules on the walls of the cylinder, and in particular on the piston. That is how the piston maintains an equilibrium level. If there were no molecules in the cylinder, the piston would sink to the bottom of the cylinder without any resistance. This explanation was relatively easy to accept. The concept of pressure was known to physicists, and the fact that a ball hitting a wall exerts a pressure or a force on the wall was well understood."

"What was even more remarkable was the finding that *temperature* is related to the motion or the *speeds* of the molecules in the gas. That discovery was not an easy pill to swallow. We perceive temperatures, hot and cold, through sensors in our skin; this sense of cold and hot seems to have nothing in common with motion. A fast-moving bullet could be very cold and a motionless ball could be very hot. Nevertheless, at the molecular level it was discovered that the temperature is nothing but a measure of the average *kinetic energy* of the atoms; the faster the molecules travel, the higher the temperature. This conclusion was reached not by *seeing* the molecules flying around in the cylinder but by indirectly inferring from the behavior of gas."

Alice was overwhelmed by what she heard. She was sure that turning the T knob would make her feel cold or hot. But now, for the first time, she finally understood that the microscopic view was quite different from the macroscopic view. She vividly recalled how the speeds of the molecules changed when she turned the temperature knob. The mystery of how the T dial worked was finally solved!

"I should also add," added the professor, "that the molecules of water that you observed do not all fly at the exact same speed. There is a *distribution* of speeds: some are slow; some are fast. It is the *average* speed of all the molecules that is related to the temperature of the gas (Fig. 4.6).

"Now, turning the P dial clockwise is equivalent — at the macroscopic level — to

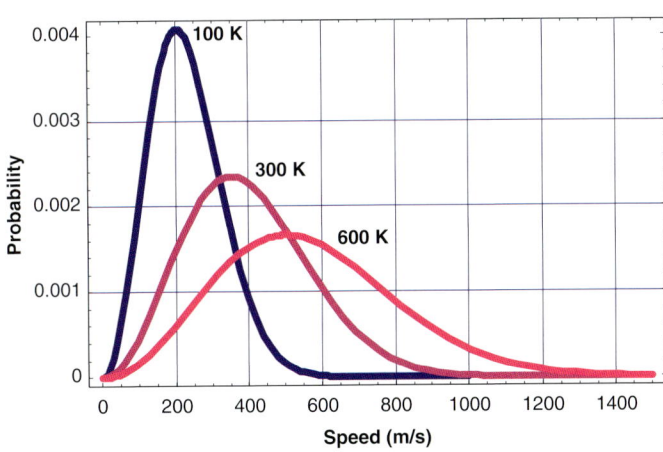

Fig. 4.6 Distribution of velocities of molecules at different temperatures.

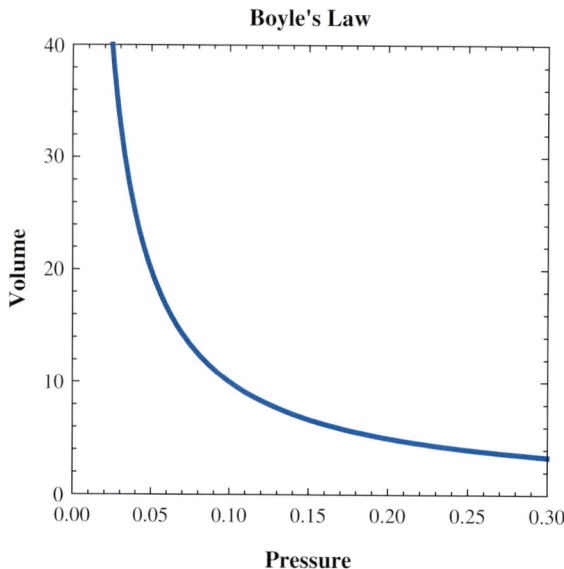

Fig. 4.7 The relationship between the volume of a gas and the pressure.

increasing the weights on the piston. At the molecular level, what you see is the molecules hitting the piston, as well as the other walls of the cylinder. The larger the pressure of the gas — keeping the temperature fixed — the denser the molecules will be, and the more frequent the collisions between the molecules and the piston."

While he was speaking, Professor Holmes changed the weights on the piston — 100 grams, 200 grams, 300 grams, and so on — and each time he added weights, the *volume* of the gas was reduced. Then he showed Alice a graph with several curves.

"*The larger the pressure P, the smaller the volume V.* This empirical law was discovered by Robert Boyle (1627–1691). What is interesting is that this law was found *experimentally*, simply by measuring the volumes of the gas for different pressures. Much later, the same relation was derived from the molecular theory of gases."

"The change in the temperature, which you have already achieved by turning the T dial, is viewed or perceived in very different ways. Macroscopically, we either heat or cool the cylinder by supplying *thermal* energy. On the other hand, this thermal energy, from the microscopic point of view, is translated into the *kinetic energy* of the molecules. The more energy we supply to the gas, the faster the motion of the molecule — faster in both translation and rotation — and if the temperature is very high, you will also observe faster vibrations within each molecule."

Noticing Alice seemed to be growing impatient, the Professor quickly recapped his lecture. "To summarize," he began.

"(1) When I add weight to the piston, the gas is compressed; the pressure increases and the volume decreases. This is true for any pure phase. The relationship between the pressure of the gas and its volume at a fixed temperature is called Boyle's Law."

"(2) Heating the system is the same as providing additional energy to the system. When the system is a pure phase, its temperature increases. However, when there are two phases at equilibrium, supplying thermal energy does not change the temperature of the system; all the energy supplied goes into transferring molecules from one phase to another. If there is only one phase and you turn the T dial clockwise, the motion of the molecules will speed up. However, if there were two phases at equilibrium, all the energy would be used to bring the molecules from the solid or liquid phase into the gaseous phase. Therefore, the speed of the molecules does not change, and hence, the temperature does not change either."

Alice felt that she fully understood what was meant by the macroscopic view — what we see and measure in the laboratory — and by the microscopic view — what occurs at the molecular level. It was something of a revelation. Alice wondered whether this was the result of her IQ boost. She could not make up her mind.

Equipped with her new ideas about temperature, energy, pressure and volume, Alice could hardly wait to get back into the professor's machine and experience the vapor phase of water again. But this time she wouldn't use the IQ machine. Her "brain power" had already been given a boost, and she was sure she would understand what was happening in the next experiment.

The next thing she knew, Alice was back in the vapor phase, running playfully on top of the molecules!

Arm-in-arm with one of the molecules, the TP gadget in the other hand, Alice positioned herself securely for a test flight. She found that even without touching the dials, the speed of her flight was constant, but after each collision the speed and direction of flight changed — at times slower and at others faster. This was what the Professor had called the distribution of velocities.

When Alice turned the P knob clockwise, she remembered that the volume of the entire gas was compressed. This meant that the density of the molecules, i.e., the number of molecules per unit volume, also increases. True enough, not only could Alice count a larger number of molecules in her immediate vicinity, but she could also feel that the number of collision per unit time had increased. Next she tried to turn the T knob clockwise, knowing that this corresponded to supplying energy to the gas. She knew she wouldn't *feel* any hotter, as the temperature, as measured from the 'outside' is expressed in the 'inside' as the mean kinetic energy of the molecules. Sure enough, as she turned the dial clockwise, all the molecules were 'energized.' It felt as if the molecules were getting an energy boost, allowing them to fly and spin faster. Turning the T knob counterclockwise towards lower temperatures, on the other hand, seemed to sap the molecules of energy — as everything soon slowed down.

With the knowledge she had acquired earlier, this new experience was far more pleasant than her first visit. Everything she observed made sense — although her senses were not involved in "sensing" the pressure or the temperature. How wonderful it was experiencing the world from the microscopic point of view! She was now more ready than ever for her next exploratory trip — into the solid phase of ice.

5
A Visit to the Solid Phase of Ice

One, two, three… The cuckoo clock's melodious chime filled Alice's room at the stroke of seven. She scurried to the window as she heard squealing from outside. A delightful sight greeted her as she drew back the curtains.

The tall trees lining her street were thickly blanketed with snow (Fig. 5.1). So heavily snow laden were they that some of the branches had broken and were dangling precariously from electricity transmission wires. The thick snow that had accumulated overnight had created a makeshift skiing area. Youngsters clad in ski gear were already deftly maneuvering their way among the crowd. Young couples with small children were having a field day in the freshly fallen snow.

Fig. 5.1 Snow.

Fig. 5.2 Father and his twins building a snowman.

Just across the street, Alice's neighbors, the Morgans, were busily constructing a snowman. The Morgan twins had two identical toy shovels and pails and were burrowing excitedly into the snow while their father carefully crafted the snowman's belly. Mrs Morgan sat on the porch, lovingly watching her husband and the children (Fig. 5.2).

Alice was so engrossed in this delightful scene that she almost forgot about her meeting with Professor Holmes that morning. She quickly got ready to go.

On the way to the laboratory she thought about the snow being white when humongous icebergs are really bluish in color. Her desire to learn more today and to visit the ice's 'interior' hurried her on her way. As she approached the laboratory, she could not escape the feeling that the mysterious girl in the picture was greeting her again. But she had no time to process the feeling, as Professor Holmes was already waiting for her.

Fig. 5.3 Snow on a trees.

"Today, we are going to visit the solid phase!" he said grandly.

"I'm ready," Alice said, stepping into the professor's magical booth, and in no time at all she was surrounded by a spellbinding world of ice.

Alice was mesmerized! There were so many fascinating and beautiful things to see and explore inside solid ice. It could not have been more different from what she had seen in the vapor phase. Here, the molecules were arranged in perfect order, laid out in a regular pattern. As in the vapor phase, the connections between the oxygen molecules were formed along the *arms*. But now almost everything had been rendered immobile. Everything was 'frozen.' There were no longer any flying or spinning molecules. She knew that the temperature was below zero, but she did not feel cold. This time, she had not expected to be cold either (Fig. 5.4).

Fig. 5.4 Alice in two views of ice.

Alice wondered where all the *kinetic energy* of the fleeting molecules had gone. She would ask the Professor when she saw him again. For now, she had work to do. She wanted to see more of the exquisite structures that surrounded her.

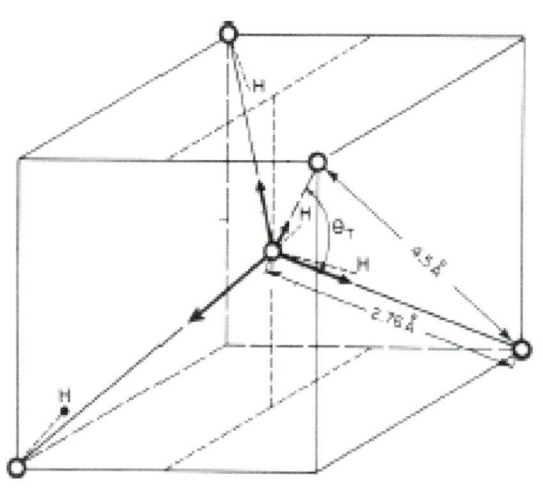

Fig. 5.5 A tetrahedral structure around an oxygen.

She hopped from one oxygen atom to another, noticing that the distance between any two neighboring atoms was almost constant (about 2.76 Å, that is, about 3/100,000,000 cm) with slight vibrational motion causing elongation and contraction of the space in between. There were two hydrogen atoms located roughly at the same distance from the oxygen as in the single water molecule in vapor (Fig. 5.5).

As she sat resting on one of the oxygen atoms, Alice looked around

A Visit to the Solid Phase of Ice 65

and saw oxygen atoms in four directions, forming a perfect tetrahedron. She also noticed, when sitting on any other oxygen, she could see exactly the same tetrahedron structure. From this vantage point, it looked like the structure of the ice was the same whichever direction she viewed it from. Yet there was something peculiar about the overall structure when she looked in different directions, but beyond the immediate four neighbors.

"How can it be," Alice thought, "that the immediate surroundings seen from each oxygen are the same, yet the overall structure is different in different directions?" She knew she would have to ask the Professor about this apparent paradox.

As she looked closely at the two hydrogen atoms belonging to the oxygen on which she sat, she noticed that the O–H distance was about the same as in the vapor molecules, but the H–O–H angle was slightly different. In a single water molecule the H–O–H angle was about 104°, but now in the solid it was 109°. Why was that?

Looking even more carefully, Alice noticed that the regularity of the oxygen pattern was such that each O–H arm was attached to an unseen arm of the neighboring molecule — an arm that was 'behind' the two O–H arms of the neighboring molecule (Figs. 5.6 and 5.7).

O—H-----O ⇌ O-----H—O

Fig. 5.6 The hopping of H.

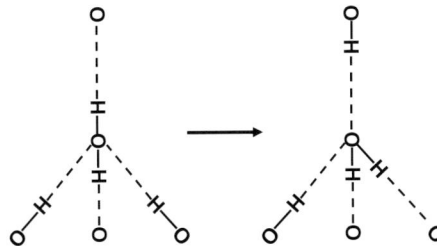

Fig. 5.7 Oxygen with four hydrogens.

While the four oxygen atoms were located almost at the same distance from the point where she sat, the O–O 'bonds' were different. Along two of the O–O bonds, she could easily see the two hydrogen atoms that 'belonged' to her oxygen atom, but along the other two O–O bonds, she saw two hydrogen atoms at a distance of about 1.7 times the O–H distance.

It was as if each oxygen atom did not have just two arms, as she had found out from the vapor phase, but rather four arms: two short arms that were similar to the two arms of the single water molecule, and two longer arms with hydrogen atoms located farther away from her that did not seem to 'belong.' Although the two pairs of arms were different in terms of the location of the hydrogen atoms, they were the same in the sense that all four arms pointed exactly to the four vertices of a regular tetrahedron.

Pondering what she had just seen — the curious difference between the two short and the two long arms — Alice wondered how it could be that what had earlier appeared as a short arm had become a long arm in a matter of a fraction of a second. And this had actually happened without altering the perfect tetrahedral scaffolding structure. Bewildered and bemused, she wanted to explore more. Did the molecules rotate in such a way that the arms changed places or did the hydrogen atoms on each atom hop from one site to the next and back again, giving the impression that the short and the long arms were exchanging roles?

Alice decided to scrutinize what was going on. The best way of doing so was to sit on the oxygen facing one arm and keep an eye on the hydrogen. Focusing her attention specifically on this arm, and on particular the hydrogen occupying it, she observed how the hydrogen atom of that arm hopped from one position to another: first it was near, and before she knew it, it had moved further away.

$$O-H \cdots O \rightarrow O \cdots H-O$$

She also noticed that while the hydrogen she had her eyes on moved away from her, another hydrogen behind her, attached to a different arm, was coming towards to 'her' oxygen. The idea dawned on her that this continuous 'dance' of hydrogen atoms hopping between two sites on the O–O line was orchestrated in such a way that *on average* for each oxygen there would be only two hydrogen atoms that are closer to it, or belong to it; and two hydrogens that are farther away and belong to a neighboring oxygen. At any given time, each O–O line was occupied by only one hydrogen, once in a while hopping from one site to the other but in such a way that no more than two hydrogen atoms were near each oxygen (Fig. 5.8).

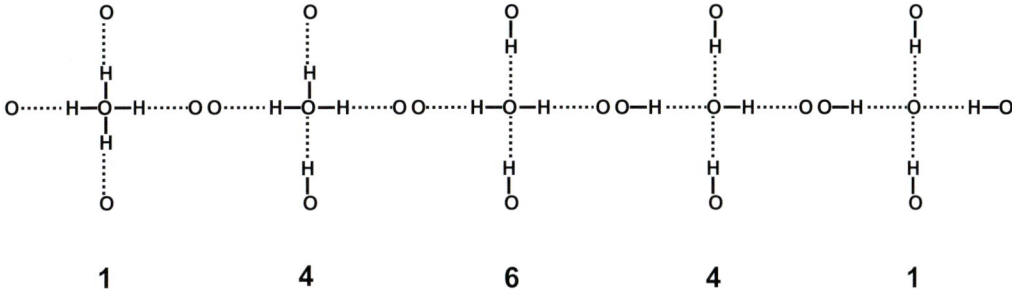

Fig. 5.8 Various arrangements around an oxygen.

Alice was fascinated by the beautiful choreographed dance of hydrogen atoms. She was certainly going to report her observations to Professor Holmes. She was almost certain that the Professor had never *noticed* that dance before. Fixing her eyes once more along the O–O directions, Alice noticed one more interesting phenomena that she thought perhaps explained the different structures observed in different directions — even though the same tetrahedral vertices occupied by four oxygen atoms could be observed from each oxygen.

Looking along one of the O–O lines, Alice saw that aside from the neighboring oxygen atom, there were three more oxygen atoms, which she named 'second-nearest-neighbors.' They were located at about 1.6 times the distance between the two nearest-neighboring oxygen atoms. Although the same view could be seen along any of the O–O directions, she noticed subtle differences in the arrangement of the three nearest-neighbors oxygen atoms relative to the three second-nearest-neighbors around her oxygen atom, depending on which direction she looked (Fig. 5.9).

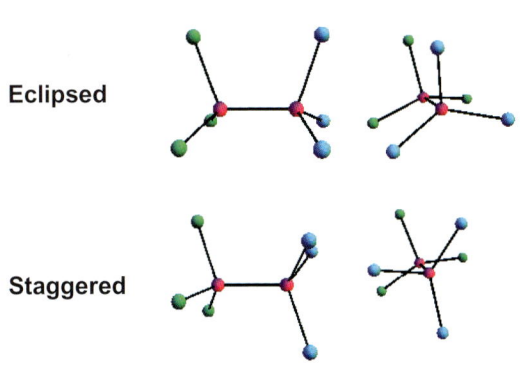

Fig. 5.9 Eclipsed and staggered configurations.

She remembered the hexagonal tiling she observed on the bottom

68   Alice's Adventures in Water-Land

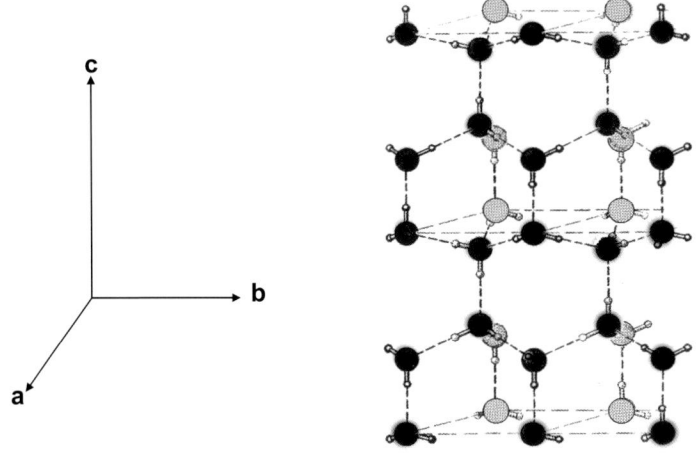

Fig. 5.10 The three axes a, b, c.

layer when ice had just started to form. She called this the 'ab plane.' Perpendicular to this ab plane was the c direction along which hexagonal layers where piled on top of one another. As she looked along the O–O bonds in the c direction, she saw that the triplet of the second-nearest neighbors was roughly in the same orientation as the triplet of nearest-neighboring oxygen atoms (Fig. 5.10).

On the other hand when she looked along the O–O line in the ab plane, she saw that the triplet of second-nearest-neighbors were rotated 60 degrees relative to the triplet of nearest-neighboring oxygen atoms. She noted the finding in her notebook and made a quick sketch. She wanted to tell Professor Holmes about it. She was sure he would would not have known it as this could only be seen from the 'inside'; Professor Holmes was obviously an 'outsider' this time and could not have seen what she had. By now, she had made a long list of observations and had an even longer list of questions to ask. Was there any indication from the 'outside' about this arrangement of the oxygen atoms? Was there any 'outside' indication of the 'hydrogen dances' along the O–O lines?

Alice suddenly realized she had completely forgotten about the professor's gadget that controlled temperature and pressure — she had been so caught up in the magnificence of the structure of the ice and all her observations!

First she fiddled with the T dial. She turned it both ways, and there seemed to be not much change. Yet, while the overall structure remained the same, small differences could be observed. Turning the dial clockwise — towards higher T — did not affect the *structure* much, but the vibrations of the oxygen atoms at each site were more discernible. Turning the T dial the other way made the vibrations less and less noticeable — until they completely ground to a halt.

There was not much to see. She knew that turning the T dial cooled or heated the ice, and that change in temperature was expressed as the vigor of the vibrational motion of the oxygen atoms. It also had a slight effect on the frequency of the hopping of the hydrogen atoms on the O–O lines, but nothing more.

Alice now turned the P dial to lower pressures, but she did not notice any effect. When she turned the dial to higher pressures, she noticed that the O–O distances became slightly shorter, thereby compressing the overall structure. However, the shortening of the O–O lines along the c axis was slightly different from that along the ab plane. She was not sure if what she saw was real or merely an illusion, but she made a note of it so she could report it to the Professor.

Just as she was thinking about leaving, Alice inadvertently turned the P knob sharply to very high pressures — and was suddenly frightened. All the regular structures collapsed! The thought that she had caused it scared her. She immediately turned the knob back to its original position and frantically yelled out the password. As she found herself back in the dark booth once again — heaving a guilty sigh of relief — she decided she would avoid reporting this particular observation to Professor Holmes. She surely must have done something that she was not supposed to do.

Exploring 'inside' was certainly fascinating, but it was much more reassuring being back in the real world. "I am back," she said to herself, feeling safe and secure in the comfort of the laboratory's familiar surroundings. She could hardly contain her excitement when she saw the Professor again. While there were only the two of them, she somehow felt a 'third presence,' that of the mysterious girl in the picture — who always seemed to be listening intently to her stories.

Alice told the professor how she saw the structure of ice, about the beautiful arrangements of the oxygen atoms at the corners of the regular tetrahedron, about the strange view along the different O–O axes, and — before the Professor could interrupt her — about the arrangements of the hydrogen atoms and the 'dance' over the O–O lines. One observation she refrained from mentioning was the structural collapse she observed when the P dial was accidentally turned up to very high pressure.

With a gentle wave of his hand, the professor finally interrupted Alice's babbling and started to explain what she had seen.

"You did an excellent job and your observations are quite significant. Indeed, the structure of ordinary ice is very beautiful. Scientists refer to this most common ice as 'I_h' — hexagonal ice I, to distinguish it from high pressure ices, denoted ice II, ice III, and so on."

"As you say, if you sit on any one oxygen atom, you will always see four other oxygen atoms at a distance of about 2.76 Å from you and located at the corners of a regular tetrahedron. In a sense, this resembles the *structure* of a diamond," the professor added, watching Alice's eyes growing bigger as she heard the word. "Yes, Alice, diamond. Aren't diamonds a girl's best friend?" he said with a chuckle."

"Diamond is a perfect crystal of pure carbon — each atom denoted C. If you had been sitting on one of *these* atoms, you would have seen the same picture around you as that of ice, but instead of four oxygen atoms at the corner of a regular tetrahedron, there would be carbon atoms."

"However, as you have observed, there is a noticeable difference in the structure of ice when viewed along different axes. This is an important difference between the structures of diamond and ice. In diamond, when you look along the C–C bond, you see that the three carbon atoms adjacent to one carbon are always rotated by 60° with respect to the second triplet of carbon atoms."

He showed her an illustration matching what she had seen.

"In ice (I_h), as you correctly observed, the arrangement of neighboring oxygen atoms differs depending on whether you are looking along the *C* axis or

along one axis in the ab plane. Scientists call the first arrangement *eclipsed* and the second one *staggered*. This difference in the *configurations* of the oxygen atoms viewed along the C axis or an axis along the ab plane is manifested in the properties of ice. For instance, the compressibility or the thermal expansion coefficient is different when measured in different directions."

"I should explain that 'compressibility' is simply the change in the volume of a given amount of substance caused by a change in the pressure. If we compress the crystal of ice along different axes, the response of the volume will be different."

"Similarly, the thermal expansion coefficient is a measure of the response of the volume of a given amount of substance to changes in the temperature. So a change of, say, one degree Celsius will cause a different change in the length of the crystal along the different axes. These facts are consistent with the structure of ice as you have observed, as well as the change in the O–O length as you turned either the T dial or the P dial."

"By the way, I should also mention that there exists an ice structure that is *identical* to that of diamond. But this ice is not stable and can only be obtained under very special conditions of temperature and pressure."

"Professor Holmes never ceases to amaze me," thought Alice. He had never even visited the ice from the 'inside,' and yet he was so knowledgeable about everything she had observed. As if reading her mind again, he continued to explain.

"You will be interested to know that scientists figured out the structure of ice not by seeing the arrangement of atoms *directly* as you have experienced but rather indirectly. They shone X-rays — the very same powerful rays invented

by Röntgen that are used in medicine today — on the crystal of ice, as well as other crystals. They analyzed rays that are *diffracted* from the crystal and using an ingenious mathematical procedure called *Fourier transform*, they 'translated' a seemingly meaningless pattern of dots from the diffracted X-rays into a very meaningful three-dimensional structure, which in the case of ice I_h was exactly what you saw from the 'inside.' Using this technique, scientists have discovered the *molecular* structures of many crystals, even of very complex molecules such as proteins and nucleic acids. So you see, you do not actually have to be *there* to be able to *know* the structure of ice. Although we cannot *see* it directly, we can infer the molecular structure from measurements performed on the macroscopic crystals."

With her mouth agape, Alice listened to all of Professor Holmes' explanations.

"How on earth did he know all this?" Alice wondered. "And I thought for once I could teach him a thing or two… Not only could he explain something he has never seen, but what is even more amazing is his capacity to explain everything so clearly and intelligently."

"I should add one more comment about the technique used to reveal the structure," Professor Holmes continued. "As I have explained to you, with X-rays we can only 'see' the structure of ice indirectly. However, with this technique one can only see the arrangement of the 'big' atoms — the structural arrangement of the oxygen atoms. This particular technique cannot 'see' the hydrogen atoms, and for a long time no one knew how the hydrogen atoms were arranged in the crystal of ice."

Hearing that X-rays cannot see the hydrogen atoms, Alice jumped triumphantly to her feet. Finally, she knew something that the Professor did not know — and perhaps no one knew. Perhaps she could teach him something new after all! She had seen, with her very own eyes, something that could not be deduced from the analysis of the X-ray diffraction patterns.

Brimming with confidence, Alice told the Professor everything that she had witnessed concerning the 'dance' of the hydrogens. With the breathless abandon of a child who has just discovered something that no one else has seen before, Alice excitedly shared the graphic details of her newfound discovery.

Professor Holmes, amused and absorbed by Alice's mastery of the things she had seen, listened patiently. Alice was wearing the smug countenance of someone awaiting an opponent's acknowledgment of defeat.

"What you have observed about the locations of the hydrogen atoms was known long ago before scientists invented new methods of 'looking' at hydrogen atoms in the crystal, the so-called neutron diffraction experiments," said the professor. "But what you have seen was known even before the

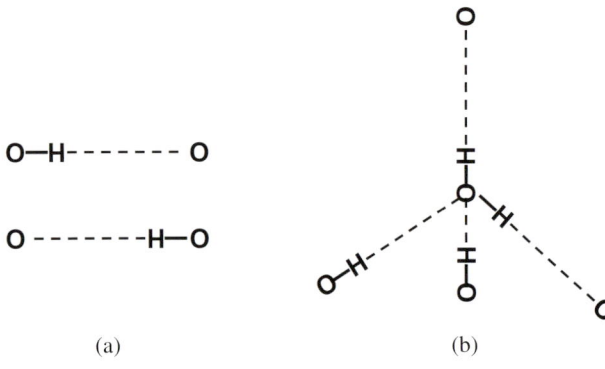

Fig. 5.11 The ice rules.

structure of ice was known. It's encapsulated in the Bernal–Fowler ice rules. These rules simply state what you have seen, namely: (1) each O–O line contains one and only one hydrogen atom, and (2) each oxygen atom has two neighboring hydrogen atoms and two distant hydrogen atoms."

"Bernal and Fowler stated these ice rules in 1933 (Fig. 5.11), even without seeing the hydrogen arrangements in the ice structure. The rules were deduced indirectly from measurements on the *macroscopic* crystals of ice. What is more amazing is that way back in 1935, Linus Pauling explained the *residual entropy* of ice based on the Bernal–Fowler rules."

Fig. 5.12 Snow on twigs.

Fig. 5.13   Glaciers, Perito Moreno, Argentina.

"Let me tell you very briefly what *residual entropy* means. We shall not need to discuss the most interesting concept of *entropy* here. At the moment I only wish to communicate to you the methodology that scientists used to infer about the microscopic world without ever visiting the 'inside' — only by measurements performed in the macroscopic world supplemented by ingenuity, hard thinking and analyzing the results of these measurements."

"For our purposes, I shall only say that the *residual entropy* is related to the *number of arrangements* of the atoms in the crystal. As you have observed, the arrangements of the oxygen atoms in space is fixed. You might have observed very small vibrations of the oxygen atoms, but aside from these small vibrations, the locations of the oxygen atoms in space are fixed. We say that there is only *one* arrangement of the oxygen atoms in the ice structure."

"On the other hand, as you have observed, the hydrogen atoms were hopping from one side to another. This means that there are many possible arrangements for the hydrogen atoms. The *number* of such arrangements is reflected in a quantity we call *residual entropy*. This is a very interesting relationship between a quantity we can *measure* in the macroscopic world, on one hand, and a quantity that cannot be *seen* by the macroscopic eyes, on the other."

"As I mentioned, Linus Pauling used the two ice rules to 'count' the total number of arrangements of hydrogen atoms in the ice crystal. It was not an exact count but surprisingly close to the exact one. The important thing is not the mere *counting* of the number of arrangements but the way this has aided in understanding that mysterious-sounding concept of *residual entropy*."

Sensing that Alice was getting exhausted by the voluminous information she was trying to absorb and process at the same time, the Professor said, "I know that you must be getting tired, but before we call it a day, I would like to say why this is such a remarkable result.

"Residual entropy is a measure of the number of possible arrangements; in our case the number of configurations of the hydrogen atoms in the ice lattice. The exact mathematical solution of the problem is extremely complicated. However, the same quantity that we call the residual entropy can be obtained by heating and cooling water from the low temperatures of ice to the high temperatures of steam. These measurements are done on the *macroscopic* water, yet they reveal some important aspects of the 'inner' structure of the water."

"I should also add that the concept of entropy gives a measure of the multiplicity of configurations. Some scientists used to refer to entropy as a measure of disorder. Remember, the perfect *order* of the oxygen atoms in ice? You cannot say that the hydrogen atoms were ordered, could you?"

"Well, this was quite a long session. I am sure you need to rest but at the same time reflect on the marvelous connection between the two 'levels' of studying nature. When you have time, please have a look at Appendix A." Professor Holmes handed Alice a few handwritten pages. "It is not very difficult and while it requires some counting, the result is interesting and illuminating."

Just as Alice was about to leave the room, Professor Holmes asked casually, "By the way, have you explored the high-pressure region in the phase diagram?"

Alice was mortified. She thought she would not have to tell the Professor about the collapse of the structure of ice that she observed when she turned up the P dial to very high pressure. She was quite certain she had done something terribly wrong and ruined the whole experiment. She would have to own up to it.

"I'm so sorry, Professor," she said glumly, after a moment's hesitation. "I did try the high pressure region, but perhaps I was too rough when I turned the P dial… I've messed up everything!"

"But before I left," she added hopefully. "I turned the dial back again and the structure of the ice reappeared…"

The Professor interrupted. "You don't need to apologize; you did nothing wrong. Indeed, when you increase the pressure to very high values, you do *destroy* the ice structure — but you get other new structures instead. Time for you to go home, Alice."

Fig. 5.14 Fishing through the ice.

She felt relieved after the Professor dismissed her. It had been a long and tiring day. Despite fatigue creeping in, her mind was awash with the images of the things she had seen 'inside.' While it was true, as the Professor had said, that the macroscopic view of ice was quite different from the microscopic view, there seemed to be no relationship between the two, or was there? She was a little confused again, but she shrugged off the thought, knowing she could very well clarify this with Professor Holmes.

Making a detour home, she passed by the lake and saw that the water was all frozen. Not unusual at this time of the year, she thought. But wait! Was there a man sitting on a chair in the middle of the frozen lake? Was that a fishing rod he was holding? The picture baffled Alice. The man was obviously fishing (Fig. 5.14). But fishing in the ice? Are there any fish between or inside water molecules? She remembered those ice tunnels in the structure of ice, and she could not reconcile the image of someone fishing *macroscopic* fish in the *microscopic* tunnels in the ice.

"Something must be amiss," Alice said to herself, too tired to do anything but shrug off the thought.

6
Alice's Visit to the Liquid Phase

Even before the cuckoo clock could do its job, Alice was awakened by what sounded like water pounding on the roof. Rubbing sleep from her eyes, she ran to the window, only to find the flowers in the planter boxes overflowing with rainwater. She was tempted to go back to bed and curl up to sleep, but how could she let a day pass without her scientific explorations? She hurriedly left for the laboratory, braving the rain to get to there on time (Fig. 6.1).

Although she had experienced almost everything that could be seen in the vapor and solid phases, Alice was not ready to explore the liquid phase. She recalled Professor Holmes' first lecture about the importance of water to life, and how interesting liquid water was, how interesting its properties, some of which she did not fully understand.

Knowing full well that what she had seen and learned so far was only the tip of the iceberg, her spirits were

Fig. 6.1 Alice walking in the rain.

Fig. 6.2 A rainbow.

buoyed by the thought that she was going to learn and explore more, as each of her sojourns provided her with a wellspring of information.

On the way to the laboratory, she was mesmerized by a beautiful rainbow that seemed to crown the cornfields, its spectacular arc matching and blending with the silhouette of the hills. As she looked closer, she realized that there was not one, but two rainbows! One had very strong colors, while the other was faint and diffused. The colors in the second rainbow were a mirror image of the first rainbow. She wondered whether this phenomenon had anything to do with water. She added this question to an increasingly long mental list of questions she planned to ask the professor (Fig. 6.2).

Alice arrived soaked to the skin, but this did not deter her from being her usual cheerful self. She could not help but look at the portrait of the young girl in the foyer. By then, Alice had gotten used to greeting her silently — even before she greeted the Professor.

"Hello Professor Holmes," Alice said brightly.

The Professor could not hide his amusement seeing Alice all drenched and disheveled. He greeted Alice with his vintage impish smile.

"Good morning Alice! You are one determined young lady. Rain or shine, you're always raring to go!"

Alice's cheeks turn scarlet at the Professor's comment.

Sensing her embarrassment, the Professor turned serious and said, "Your timing is perfect because we are moving on to the next stage of our exploration: the liquid phase."

"I could not let the day pass regardless of the bad weather. I know there's so much more to learn and explore. I am eager to see the liquid phase from the inside," Alice said imploringly."

"I believe it is not going to be that simple," replied the professor. "If you think going into the vapor phase was a breeze, that is because it is relatively simple since most of the time the molecules were single and detached. In the scientist's parlance, we say that the molecules in the vapor phase are *independent* of one another, and although at times there are collisions, these moments of *interactions* are very rare. Therefore, one can study, as you have done, the single water molecule without bothering about the other molecules."

"The other extreme case, which is the solid phase, is also relatively simple. In this case, the molecules do *interact* with each other, and therefore we say that they are *not independent*. As you have noticed, each molecule is almost permanently bound to four other molecules in its neighborhood, but because of the regularity of the structure of the solid it is relatively easy to record the structure and to understand its behavior."

"I should also add that scientists of the 19th century have succeeded in understanding the *macroscopic* properties of gases from the *microscopic* properties of the molecules comprising that specific gas. This was a great achievement of the molecular theory of matter. Later, by the early 20th century when techniques were developed to "*see*" the structure of solids, the theory of the solid state had also fully ripened. However, the theory of the liquid state, even for simple liquids such as argon, has lagged behind."

"The liquid phase is notoriously difficult to fathom and describe. Even simple liquids consisting of inert gas atoms such as neon or argon pose serious theoretical challenges. This is *a fortiori* true for liquid water, which is a very complicated liquid. We shall explore this difficult, yet very fascinating phase in the next few sessions. Are you ready to dive into the microscopic world of liquid water?"

Teetering on the flat edge of the rocky cliff, Alice balanced herself and tried to concentrate on the deep blue water awaiting her when she made her dive. The water looked treacherous but her adrenalin was pumping so fast she could hardly wait to plunge into the waves. With one deep breath she dove from the cliff as her heart thumped against her chest. She felt the cool wind in her hair. She was just inches from the water surface when… (Fig. 6.3).

"Alice, are you with me? Your mind seems to have wandered elsewhere," the professor said with a chuckle. "I have asked you a question but you did not seem to hear."

"I am so sorry," Alice replied with crimson cheeks hot with embarrassment. "I heard your question, but I was imagining myself diving off a cliff into the deep blue sea… The thought of diving into liquid water conjured up some beautiful images of what lies beneath."

"I understand," the Professor said. "Exploring something new can be a beautiful experience indeed. The newness probably makes it a little scary and picturing pleasant experiences might ease the fear that goes with trying something out of the ordinary. But as I was saying — before you took the plunge — I strongly suggest that you go and 'recharge your batteries' in the IQ machine. That will enhance your ability to comprehend the liquid phase."

All of Alice's doubts evaporated upon hearing the Professor's reassurances.

"Take a deep breath before you dive into the liquid water. Watch carefully and if you find something strange or something interesting, make a note, and we shall discuss it later," the Professor added.

Alice was so anxious to start her expedition that she hardly allowed the Professor to finish his sentence. She agreed to the professor's suggestion to get a brain boost in IQ machine

Fig. 6.3 Alice dreaming of diving into the blue water.

first. This trip was not going to be as easy as the previous excursions in the vapor and the solid phase.

As before, she did not feel herself getting any 'cleverer.' But she trusted Professor Holmes and she was confident that he knew exactly what he was doing — and subjecting her to. After her brain boost, she immediately got into the shrinking machine and got herself ready to 'plunge' into the liquid phase. As in her previous forays into the unknown, she kept her eyes shut for a moment, afraid that she might see something she was not prepared for. But each time she had opened her eyes, what greeted her was a delightful new experience.

This time Alice saw a sort of mixture of the vapor and the solid phase. The molecules were not at fixed lattice points as they were in the solid phase, but were highly motile as in the vapor phase. She noticed, however, that the density of the molecules was much higher, and the collisions between the molecules were more frequent. Sometimes aggregates of two, three or four molecules formed and stayed together briefly, only to separate from each other and then travel in different directions, forming new aggregates with other molecules. This process of flickering aggregation went on and on with no sign that any permanent *structure* would settle (Fig. 6.4).

Alice's first impression was that of a complete *disorder* — no regularity and no structure — which persisted for a long duration. So chaotic was it that she did not even know how to describe it. She could not continue to wander in this total *balagan*, a word she frequently heard the professor uttering when situations got messy.

Fig. 6.4 ALice in the liquid phase.

Iguacu Falls.

She decided to check systematically how this disorder came about from either the vapor phase or from the solid phase. Perhaps that would provide her with some clues about how to understand this highly confusing system. At that moment she felt that she had benefitted from the effects of the IQ machine. She knew that if she had not used the machine, she would not even dare to undertake the task of studying this extremely complicated, nearly chaotic system.

She remembered the point Z that was marked on the phase map of water. That was the point were the vapor coexisted with liquid. To the 'east' of point Z, she knew there was only one vapor phase, so she turned both the T and the P dials and navigated herself to point Z. Slightly beyond that to the east, she saw the familiar landscape of the vapor phase, where the molecules flew at random with occasional collisions.

Turning the T dial counter-clockwise, towards low temperatures, she saw that clusters of molecules were formed — first a cluster of two, and then threes and fours, and so on. She noticed that the clustering of molecules did not last long. They formed by joining arms, stayed together for short periods of time and then separated again, and each molecule went on its own way and sought out new partners to form new clusters. It dawned on her that perhaps the water molecules were plotting some conspiracy! Were they gathering into small groups, whispering messages, and then parting ways so that the message could be spread around to other molecules?

The lower the temperature, the larger the average cluster size became. But at some point the cluster size became so large that they could be called droplets of liquid water. These droplets sank to the floor and formed the liquid phase (Fig. 6.5).

Turning the T dial further did not produce any dramatic changes. As she tried turning the dial further, Alice noticed that an increasing number of molecules sank down from the vapor into the liquid, making it difficult to tell the real difference between the two phases. Both seemed to be totally disordered. The difference, though, was only in terms of the *average density* of water molecules. While one phase was relatively dilute, the second seemed to be jam-packed.

Fig. 6.5 Alice observing the liquid and solid phases.

Next, Alice decided to examine the liquid from the solid perspective. She navigated herself quickly towards point Y in the phase diagram, where solid coexisted with liquid. She decided to start at a point just 'west' of point Y — from the solid ice at a temperature slightly lower than point Y in the phase diagram.

Here, she found herself again in the familiar landscape of the perfect or nearly perfect structure of ice. Being back gave Alice a feeling of comfort and reassurance. She turned the T knob slowly clockwise so that she could observe how the *phase transition* occurred. There was a slight tremor, and the molecules started to shake. Undaunted by what had occurred, she concentrated on the other molecules in her vicinity. She noticed that from time to time one or other of the oxygen molecules that occupied the vertex of the tetrahedron also shook, occasionally tumbling down and eventually detached himself from the regular structure of ice.

As Alice turned the T knob even further, the motion of the molecules surrounding her grew increasingly violent. Although she could still see on average four oxygen atoms occupying a regular tetrahedron, it was neither regular nor static like the stable structure she had experienced in the solid phase. The more she turned the T knob clockwise, the fuzzier the images around her became, until no recognizable structure could be discerned. No longer did the exact four nearest neighbors occupy the vertices of the tetrahedron. Instead, the space around her became increasingly overcrowded so that at times she observed five or even more molecules coming closer.

She also noticed that those regions where the structural breakdown occurred sank to the bottom, while at the top she could still see the orderly arrangement of the ice molecules. From a distance, it was clear that there were two phases, one on top of the other. The ordered phase — the solid phase — was on top, and every patch of broken-structured aggregate sank to the lower level — the liquid phase.

As she was observing the solid ice on top of the liquid water, the strange scene she had witnessed that morning came to her mind: a man sitting on the frozen lake and threading a baited string through a hole in the ice. She was

Ouzoud Falls, Morocco.

certain that this scene was the same as what she was now observing — from the inside!

But she had to continue her exploration of the liquid phase.

As she continued to turn the T dial, more and more ice melted. The structure disintegrated and sank down until all of the structured phase had completely disappeared.

Although it was difficult to describe the scene around her, one thing was perfectly clear: the *perfect order* that was so dominant in the solid phase had gone. Occasionally, there were four water molecules at the four corners of the tetrahedron, but in addition to their motion, other molecules were seen invading those cavities that were totally empty in the ice. More and more of the tunnels and cavities that Alice had clearly observed in the ice were now crammed with water molecules, which seemed to defy any order or regularity.

Once all the molecules comprising the ice-phase melted, the qualitative appearance of the scene did not change much as Alice turned the T knob further. However, the extent of disorder gradually increased and the original tetrahedral structure became fuzzy, occasionally unrecognizable. The molecules were pushing harder and more frequently as if all of them were hurrying towards some unknown destination. The scene was total chaos.

"Is there any sort of regularity in this very irregular landscape?" Alice wondered, realizing that the question was self-contradictory even though that was exactly what she had observed. Then she remembered what the Professor had told her: that liquids, in general, are highly *disordered*, but that water maintains some degree of order through the entire liquid range (at 1 atm pressure and between 0°C to 100°C).

There must be some reconciliation between the regularity imposed by the strong bonds between the molecules, and the tendency towards the total irregularity caused by the random motion of the molecules, Alice thought. Just like doing a jigsaw puzzle, Alice knew she had to find two missing pieces in order to get a clear picture. This would allow her to reconcile the contradicting information about the liquid phase that she had obtained. Vague as to what she was aiming for, Alice decided she would do some measurements, hoping that the results would somehow give her some clues to her puzzle.

Knowing full well that the scene before her was but a fleeting moment and would change in a blink of an eye, she thought it best to measure some *average* quantities. Intuitively, she felt that the simplest measure was to count the average number of water molecules in some spherical spaces around her. This turned out to be a sound decision. She drew a set of imaginary spheres of different radii around her and started counting the number of water molecules that visited that sphere within seconds of one another.

Judging from the figures and from what she actually observed, Alice noticed that when the radius of the sphere became larger than or about 2.8Å, most of the time she was able to count four atoms of oxygen in that sphere. This was exactly what she had observed in the ice phase. In contrast, in the liquid phase, there were some deviations from that exact number. Infrequently, a fifth atom would enter into the sphere and even more rarely, a sixth or more counts.

Wanting to put together a substantial report, Alice meticulously and systematically examined the frequencies of the number of oxygen molecules in a sphere of radius 3.5Å, taking many snapshots of the scene around her, counting the number of oxygen atoms that were found inside that sphere — but excluding

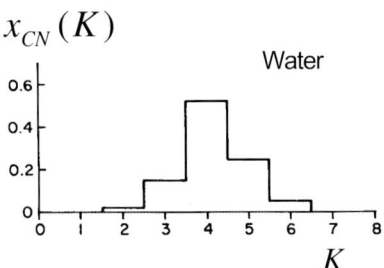 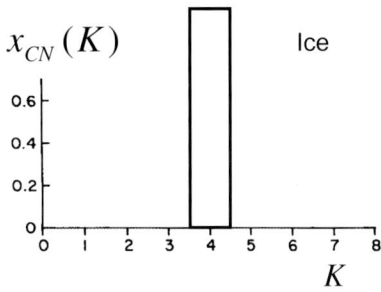

Fig. 6.6 Histogram of coordination numbers in water and ice.

the oxygen atom she was sitting on. Satisfied with her work, she then made a histogram, a plot of the occurrence of various frequencies (Fig. 6.6).

Repeating the same measurements at different temperatures as she had done previously, she noticed that the general appearance of the histogram did not change much although it appeared that the whole histogram was veering gradually to the right. As the temperature increased, the average occupancy of the sphere of radius 3.5Å increased — the *maximum coordination number* increased. The spread of the molecules over the various coordination numbers also increased.

Alice was convinced and felt confident that she had gathered enough 'new' data worth discussing with Professor Holmes. With these thoughts fresh in her mind, she yelled the password and before she knew it, she was back with the professor once again.

Beaming with pride, she showed the Professor all the tables and charts she had painstakingly put together. Her report was a clear demonstration of her thoroughness and attention to detail — something that did not escape the Professor's attention.

"You've done a very good job Alice," said the professor, clearly impressed. "This is an important start. What you have discovered is what scientists call the *distribution of coordination number*. This is exactly what you have recorded in your graph of the frequencies of occurrence of different numbers of 'neighbors,' which is also referred to as *coordination numbers*. As you may well remember, the coordination number in ice was fixed."

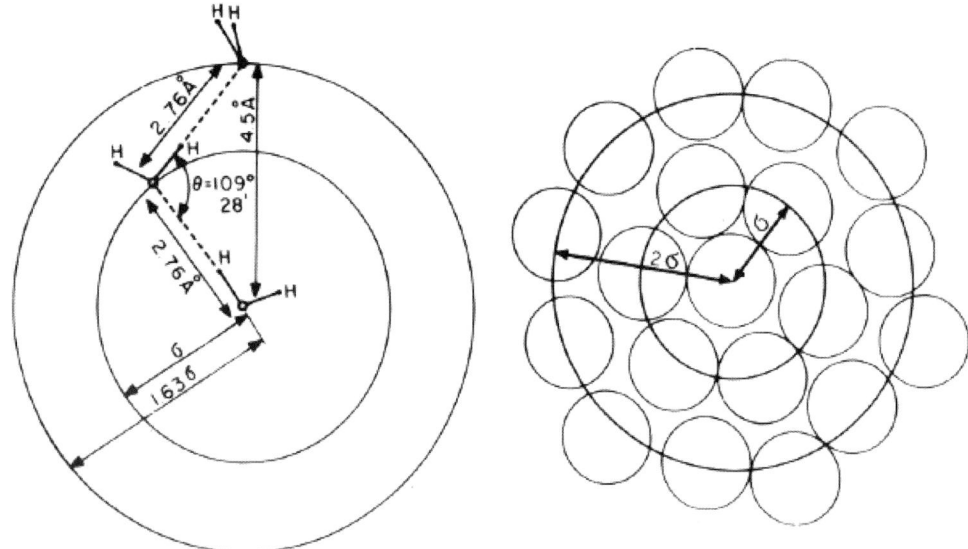

Fig. 6.7 Concentric sphere around a molecule of normal liquid (right) and around a water molecule (left).

"Yes! In ice it was exactly four, and there were no other coordination numbers," interrupted Alice, overwhelmed by the Professor's acknowledgment of a job well done.

"You are right, Alice," the Professor said. "In a solid, any solid for that matter, the coordination number is fixed. In most cases, however, that fixed number is larger than four. What you have found is that frequently liquid water has a coordination number of nearly four. This fact is very important in understanding the outstanding properties of liquid water."

"However, before discussing the properties of water, you should do some more measurements making use of the same technique, but this time an even more detailed study. If you plot the average number of molecules in each sphere (around a given molecule), you find a graph of the kind you have drawn. Note that the coordination number increases with the radius. More specifically, we know that the volume of the sphere of radius R is simply $\frac{4\pi R^3}{3}$ (or about $\frac{4 \times 3.14 \times R \times R \times R}{3}$)." (Fig. 6.8)

"Now, if we plot the *volume* of the spheres of different radii, we find that it has a similar form to the curve you drew of the coordination number. This is understandable. If you take larger spheres, you expect to find larger numbers of molecules in these spheres. It is difficult to see from such data any pattern that can be recognized as 'structure.' Therefore, I have prepared for you Appendix B, referred to as *radial distribution function*. You can also find this at *ariehbennaim.com* if you wish. You can use it whenever you want to study the 'structure' of water in more detail. Instead of counting the number of oxygen atoms in a *sphere* of radius R, you should do the counting in *spherical shells* of radius R and some width, say 0.1 Å. Also, you should compare the counting

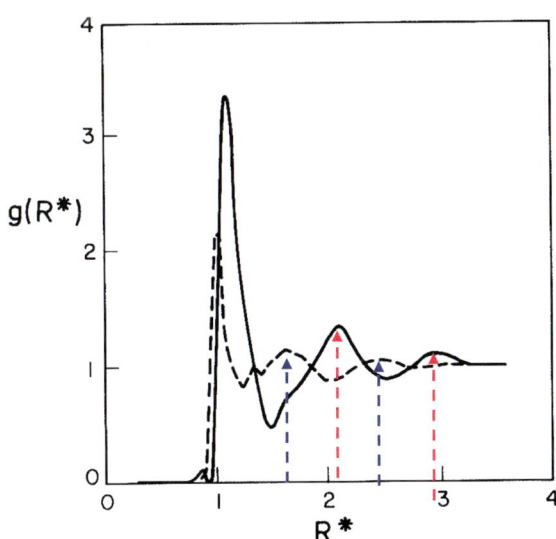

Fig. 6.8 Radial distribution function; argon (full line), water (dashed line).

do with a 'reference liquid' composed of simple atoms, like argon. By comparing the two liquids you will discover not only what is meant by the 'structure of water' but also some fundamental clues to understanding the properties of water. You have a lot of work to do in order to process and digest what you have seen today."

The Professor handed Alice a few pages. Alice only saw the title: *radial distribution function of water*.

"Why don't you go home and rest now?" Professor Holmes said. "Tomorrow is going to be another long exploratory day!"

Alice could hardly wait for the chance to find out about the *radial distribution function of water*. She took the pages the Professor gave her and kept them in her purse. Although she did not have the slightest clue as to what it meant, and what it had to do with her findings about the coordination number, she still looked forward to learning about it with the same eagerness she had when she started her excursions and explorations. The word 'distribution' in relation to water was impressive enough and this excited her all the more. She decided to study these pages as soon as she got home.

Humming a tune to herself, Alice meandered gaily back home. Although she was physically tired from the long day, her mind was active and she could not help but think of tomorrow's promise, the prospect of new doors opening to a wider span of knowledge.

"*Radial distribution function*, what a fancy and impressive term," she said to herself, repeating the line as if in a trance. Looking up at the sky, she realized that she had forgotten to ask the Professor about the glorious rainbow she had seen. But before she had time to process the thought, she was a few meters from her front door and the familiar savory aroma of her mother's roast beef assaulted her nostrils.

It was perfect day for Alice: her meticulously conducted study, Professor Holmes' enthusiastic endorsement of her work, and now coming home to delicious home-cooked roast beef. The mere thought of her teeth sinking into those succulent slices of roast beef generously bathed in rich mushroom gravy with a

Fig. 6.9 Animals on the lake.

heap of baked baby potatoes made her mouth water. And whenever her mother prepared her 'to-die-for' roast beef, she would always make a black forest cake for dessert, rich and moist and generously topped with maraschino cherries. What a rewarding day indeed and what a way to end it!

7
Why Does Ice Float on Water?

Since she was a child, Alice would spend winter breaks at her grandparents' farm. But for the very first time, she had decided to stay to forego the farm visit, opting instead to discuss with the Professor her recent explorations.

Alice's mind wandered to the beckoning of winter on the farm with its promise of panoramic views of snow-capped mountains and the lake that would become a skating rink as the temperature fell below freezing. In her childhood, she had often wondered what it was that supported the ice on top of the water. That very same lake was the scene of so many beautiful and memorable family gatherings. They had built bonfires by the lake, and huddled around a pyre of dried leaves and branches, they had taken turns roasting marinated beef ribs, lamb cutlets, hotdogs and marshmallows.

A distant wailing siren brought Alice back to reality, and her reminiscences about the good old days faded. She smiled. Her experiences of the

Fig. 7.1 Icebergs flotting on the water.

ice and liquid water would surely help her with the question she had long grappled with: "Why does ice float on water?" Besides, she could always turn to Professor Holmes for an answer.

Exhausted from cleaning her room and putting away her summer clothes, she felt a pang of thirst. "A tall glass of ice cold water will do the trick," she thought. She opened the fridge only to find that there was no cold water — and no ice cubes. She quickly filled a bottle with water and put it in the freezer.

"I'll get back to you in a few minutes," she said to the bottle. "You should be cold by then."

Putting a couple of satin throw pillows on her lap, she settled on her favorite couch, rested her legs comfortably on the ottoman, and reached out for the TV's remote control. Zapping through the channels she chanced upon her favorite TV program. "Perfect," she said aloud.

Suddenly, in the middle of her program, there was a loud popping sound from the kitchen. She rushed to the kitchen to find out where the sound had come from. She had been so engrossed in the TV program that she had completely forgotten about the bottle of water in the freezer! (Fig. 7.2).

Fig. 7.2 Bottle of water in the freezer.

"Oh no!" Alice said as she opened the freezer, gasping at the mess inside. The bottle was no more. Tiny shards of glass were scattered all over the freezer. Only the water that she had put inside the bottle was intact — frozen and taking the shape of the bottle. Although she could not have cold water, Alice thought, she could use the ice bottle instead. She opened one of the kitchen drawers, found an ice pick and carefully started chipping away at the ice. As she watched the pieces of ice floating on top of the water, she remembered the ice layer that always floated on the lake's surface. But what about the explosion? What could have caused it? Did ice contain some explosive like dynamite? Did ice on the surface of a lake also explode?

"All of these phenomena must make sense after my observations of ice and water," Alice thought. "I should be able to use all the information I have gathered to my advantage. It's amazing how much I have learned in such a short time! Thanks to Professor Holmes, all of my questions have begun to crystallize into solid concepts. His patience and guidance every step of the way have helped me obtain substantial findings in all my experiments." Then, as if some reality check bulb had just lit up, she wondered whether this 'acquired knowledge' could only have come from the IQ machine.

Alice vividly recalled the ice structure, the water molecules in spatially arranged order, leaving empty holes and tunnels in between. She also remembered that once ice melted, some of the water molecules left their lattice points on the perfect crystal and penetrated the empty holes. Ice, she recalled, had a coordination number that was *exactly* four. On the other hand, she also knew that the average coordination number of water was *larger* than four.

With this information, she could deduce that a larger coordination number meant that more water molecules were packed in a given spherical volume of liquid than ice, which meant that the *weight* of the sphere of water was *larger* than the weight of the same sphere of ice. What was true for a tiny sphere around a given molecule in the *microscopic* world must also be true in the *macroscopic* world. This meant that any piece of ice of the same volume weighed less than the same volume of water — which explained why ice floats on water. Simply put, the less dense of the two phases float.

What about the exploding bottle? Although the two phenomena seem to be unrelated, were they not both simply the result of the *open* structure of ice? This was the reason why ice floats on water. But this was also the reason why the water-filled bottle — as the water inside froze — needed more space. When the bottle did not have enough room for the ice... then... boom!

This was exactly what Alice had seen when crossing from the ice phase to the liquid phase: more water molecules entered the empty holes and tunnels, which made water denser (in molecules per unit volume, or weight per given unit of volume). This is the reason why the volume of ice contracts when it melts into water.

Clearly, the flipside is the freezing of water into ice. When water freezes into ice, all of the water molecules that occupied the holes and the tunnels form an orderly arrangement, obediently returning to their lattice points to form a perfect *open* structure of ice. An open structure means a larger volume formed by the same number of water molecules.

What happens though if the water molecules are confined to the boundaries of the bottles? When water freezes, the molecules expand, and being in a confined space within the bottle, they press hard on the walls until the bottle cannot contain them any longer nor stand any more pressure. At this point the walls collapse to make room for the expanding water molecules. What follows almost instantaneously is an explosion.

All that made sense to Alice but some doubts still crept into her mind. Perhaps she was wrong. Perhaps she was only deluding herself that she understood. "This cannot wait. I have got to know if I'm right," Alice thought. She decided to go and see Professor Holmes.

She greeted the Professor as she arrived at the laboratory, immediately telling him about what she had just experienced and coupling her story with her explanation. She wanted the Professor to confirm her reasoning.

"You're right, Alice," the Professor said after she had finished her litany of theories and ideas. "What you have just seen is one of the anomalies of water. In most liquids, the volume *shrinks* when the liquid freezes. If you fill a bottle with alcohol, or benzene, or even oil and put it inside the freezer, you will observe that as the liquid freezes, the volume occupied by the same amount of substance gets smaller as it changes into the solid phase.

"Water is very different. As you have observed, the packing of water molecules in ice is quite spacious. This is what is called an *open structure*. When ice melts, it becomes denser, meaning there are more molecules per given unit of volume."

"Do you remember the solid liquid (SL) equilibrium line in the phase diagram? It has a negative slope for water, but positive slopes for all other substances. As I mentioned in my lecture, the slope is determined by the volume change on melting. Now, you can also understand the unique negative slope of the SL line in the phase diagram of water."

After a pause, he added, "Although ice has a more 'open structure' than water, liquid water is also considered to have an open structure. We shall study this phenomenon later. But that is the reason why ice floats on a lake. As simple as it may sound, it is also a very crucial factor for the evolution of life in the seas and the lakes."

"Huh? Whose lives was the Professor talking about?" Alice was perplexed.

As ever, the Professor seemed to have read her mind and said, "Think about it. If water were a 'normal' liquid, becoming heavier as it freezes, then in winter water would freeze and sink to the bottom of the lake. In the daytime, although the sun heated the atmosphere, it would only be able to heat the lake's surface. The heating would only happen at the surface and would not reach the lakebed. At nighttime another layer of ice would form and would again sink to the bottom. The process would repeat itself night after night until all the frozen water accumulated on the lakebed — until finally what you would have is a humongous piece of ice!

"The ice that floats on the surface of water acts as a protective insulating layer. The water underneath the ice is kept at a temperature slightly above freezing, allowing for the sustenance of life forms under the layer of ice. Amazingly, a thriving aquatic community can exist beneath the ice, so perfectly adapted to their surroundings that they manage to survive. At night a layer of ice is formed on the surface, while in the daytime the heat from the sun's radiation first melts the ice on the surface, allowing for the life forms underneath to flourish."

"Some people believe that water was designed in such a way that it possesses just the 'right' properties to support life. In the course of our discussions, we shall see that water also possesses many more properties that are vital to life — at least the kind of life that we are familiar with on our planet."

Alice was beaming from ear-to-ear as she walked home. She felt a deep sense of pride at being able to put together all the pieces of the jigsaw puzzle connecting two seemingly unrelated phenomena — the ice floating on the water and the explosion of the water-filled bottle in the freezer. She did not grasp the relevance of this for life, but the very understanding of these phenomena filled her with satisfaction.

Still, at the back of her mind, she could not helping thinking that all this was a result of the IQ boost she had gotten in the laboratory the other day, and that she was bound to lose her grip on the understanding of the phenomena.

"One day, I must find out about the IQ machine," she thought. "I need to know whether its effect is temporary. Was it like the shrinking machine where one shrinks and then returns to normal again? Or was the IQ machine's effect permanent?"

Alice decided to make a little detour on her way home and passed by the lake. While she watched some kids enjoying ice-skating, it dawned on her that the thought had never occurred to her as to why ice was so slippery — thus

Fig. 7.3 Ice-skating.

making ice-skating possible. Why could one not go skating on any other solid? Was this related to the unusual properties of ice? Alice made a mental note to ask Professor Holmes about it.

As she sat enjoying the view, she saw a man not far away from the skaters, rod in hand and with his fishing line threaded through a hole in the ice. The site of a man fishing no longer surprised her. Now she understood that the ice floating on top allowed the marine life below to survive — at least before ending up on the fisherman's dinner table.

8

The Outstanding Temperature Dependence of the Volume of Water

Although Alice had understood the differences between ice and water, she wondered what Professor Holmes meant about water also having an 'open structure.' The concept was not clear. She vividly recalled that when she crossed from the solid phase to the liquid phase, all the 'structure' of the ice had collapsed. Did the word 'structure' have different meanings when applied to the solid and the liquid phases?

Today's task for Alice was to conduct a very crucial experiment involving liquid water. Professor Holmes had done a good job of preparing Alice for discovering one of the most unusual, and perhaps also most unique, properties of water. Her assignment called for some explorations from the 'interior' as well as from the 'exterior.' She knew that today's assignment would clear the cobwebs as far as grappling with the issue of open structure was concerned. Which was more 'open-structured' — water or ice? Why was water also considered an 'open structure'?

To get to Professor Holmes' laboratory, Alice had to walk through well-manicured gardens with rows and rows of neatly trimmed shaped hedges. In the middle of the garden was a fountain set with lovely sculptures of cherubs at play. On the topmost layer of the fountain were several cherubs pouring water from stone pitchers to layers below where more cherubs nestled on ledges. This pretty fountain with its innocent-looking cherubs always caught Alice's attention.

The fountain water cascaded down to a pond beneath brimming with water plants of various shapes and sizes. One particular plant fascinated Alice, though she did not know the name of it. In the daytime when the sun was shining in all its glory, she had observed that the flower would fully open and she enjoyed caressing the velvet-like petals. At dusk, however, the petals would shyly fold and close tightly, reducing the flower to a mere bulb — only to unfurl the next day. Every time she wandered past this garden, she would say to herself, "life abounds with beauty."

"Good morning Professor Holmes," Alice said in her chirpy voice. She had already given the girl in the picture her usual silent greeting. It had almost become a routine by then.

"Are you ready for today's task?" the Professor asked.

Alice saw that the Professor had set up two experiments. They seemed to be quite simple. The first apparatus had a flask of water with a scale where one could accurately read the *volume* of liquid. In the second set-up, the flask was filled with alcohol (more precisely, methyl alcohol, or methanol) (Fig. 8.1).

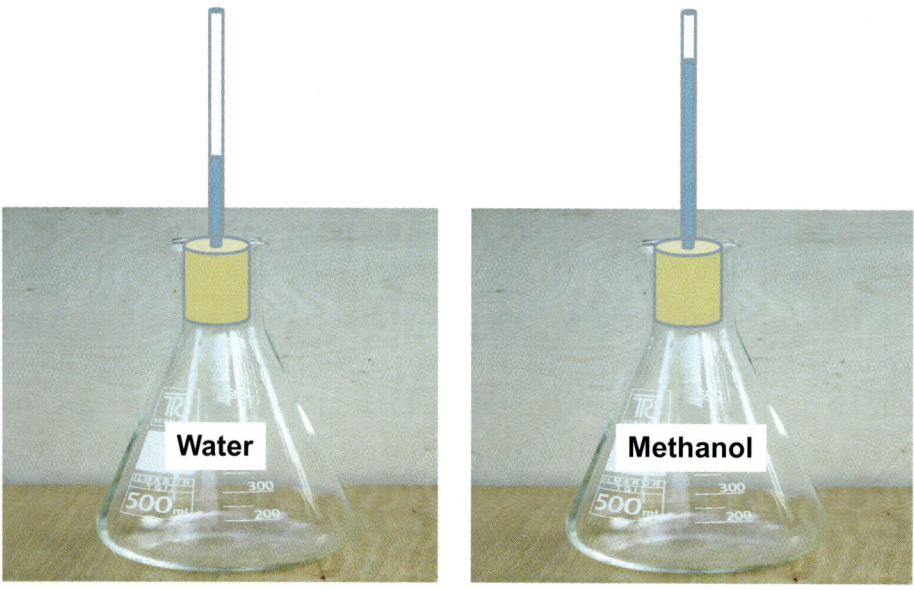

Fig. 8.1 Measuring the volume of water and methanol.

The two flasks were initially filled with the two liquids at the same temperature (0°C), and at the same pressure (1 atm). The whole experiment was conducted in a well-ventilated chamber.

"Water is of course not a 'dangerous' liquid," Professor Holmes had commented, "but methanol is poisonous, so we have to be extra careful not to inhale the vapor of methanol as large quantities can cause blindness." Alice remembered that alcohol was an ingredient in all 'alcoholic' beverages, and as far as she knew no one ever mentioned that alcohol can cause serious damage — but that was not part of today's experiment.

All Alice had to do was to record the volume of the liquid at different temperatures. Professor Holmes had prepared a few sheets of paper where she was supposed to record the data.

"Today we will be doing simple experiments: measurements of the volume of the liquids at different temperatures," said the professor. "This experiment is done from the 'exterior,' but the understanding of the results is possible only from the 'interior,' i.e., from the microscopic view. Let us start the experiment now."

With that, the professor turned on the heater and Alice started recording the readings of both the temperature and the volume in each of the flasks. She paid little attention to the figures she was jotting down and did not even try to understand what they meant. She just went on recording as the temperatures of the liquids rose. The whole experiment did not last long, and once the temperature had reached about 50°C, the Professor turned off the heater. Impatient and a little irritated as to why she had to go through all those numbers and tables, Alice heaved a sigh that was not lost on the Professor.

"Now take a look at the data you have recorded. Do you see some patterns or something unusual?" the Professor asked. Alice looked closely at the figures, but there seemed to be nothing special or noteworthy.

"First, we have to *plot* the results," the professor suggested. "This is a much better way of 'seeing' a new trend or a difference in the behavior of water."

Alice was about to open her mouth when the Professor continued, in a more commanding tone, "Why don't you plot the data — the volume on one axis and the temperature on the other?"

With another heavy sigh, Alice did as she was told — without expecting any revelations. Yet as she started to plot the data, she noticed a difference between the graphs for water and for methanol. The curve for methanol (as well as all other liquids) had a positive slope, meaning that increasing the temperature resulted in an *increase* in the volume. But water, on the other hand, exhibited a quite different behavior: between 0°C and 4°C the volume *decreased* and the slope of the curve was *negative*, at 4°C the curve had a minimum, and beyond that the volume climbed upwards as for all the other liquids (Fig. 8.2).

"You see," said the professor, sensing Alice's confusion, "just merely looking at voluminous data can be confusing and one can easily get lost in the numbers. 'One cannot see the forest for the trees,' as it is commonly said. More often than not, in science, when one plots the data as you just did, important patterns immediately pop up. You can clearly see that the volume of water changes in a different way when you increase the temperature."

"What you have just seen is one of the most outstanding properties of water that renders it a unique liquid. All other liquids, as well as most materials, solids or gases, *expand* on heating. Their volumes become larger as the temperature

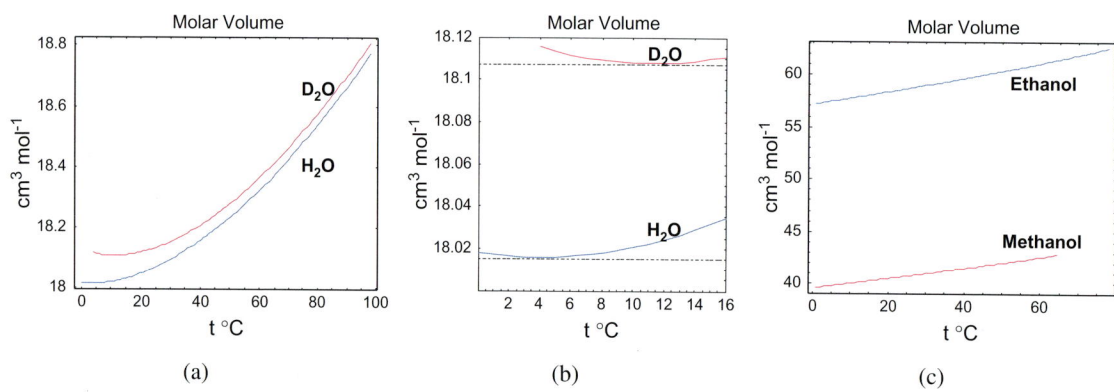

Fig. 8.2 Specific volume of water, heavy water methnaol and ethanol at different temperatures.

increases. Liquid water behaves differently in that over the range of temperatures between 0°C and 4°C, the volume actually shrinks as we increase the temperature. I should also add that there exists another form of water called 'heavy water,' denoted D_2O (instead of H_2O), which has the same behavior although over a larger range between 0°C and 11°C." The professor picked up a book from the shelf and showed her a graph for heavy water.

"To understand this better," he went on, "you will have to see what happens when you heat the liquids from the 'interior,' and then you will be able to understand why. It will also reveal one of the 'molecular secrets' of water, which explains other properties of water as well as aqueous solutions. You will also understand why any liquid — or solid or gas — *expands* when heated."

"Would that also explain why the days are longer in summer when the weather is hot," Alice interrupted, "and why they are shorter when it is cold?" Alice was sure that the length of the day behaves 'normally' like normal liquids.

"No," the Professor replied. "The length of the day has nothing to do with the expansion of normal liquids. This is quite a different phenomenon. It has to do with the orientation of the earth's axis towards the sun. It is a good comment though because it shows that you are analyzing the correlations between different phenomena."

"Now let us repeat the same experiments from the 'inside.' You will have control over the temperature using the T dial, and you can change the liquid by pressing the buttons labeled *water* or *alcohol*. This is particularly interesting because you will see what accounts for different behaviors from the molecular point of view."

Without hesitation, Alice stepped into the shrinking machine, and she was immediately plunged into the liquid. The initial temperature was set at 0°C. She first examined the behavior of alcohol, which was supposed to represent a 'normal' liquid.

Looking around, Alice saw nothing even remotely close to the tetrahedral arrangements in liquid water. The alcohol molecule — in this case methanol with one carbon atom, one oxygen, and four hydrogens (CH_3OH) — was big compared to the water molecule. The alcohol molecule from which she was observing was crowded by other molecules, with six, seven or more neighboring molecules moving around, almost all within contact distance of her molecule.

As she turned the T dial, as she had expected from her previous observations, the movement of the molecules became increasingly frenetic and they collided with one another more vigorously. She also noticed that the average distance between the molecules became increasingly large as she raised the temperature.

That was 'normal' and in a sense, Alice had expected that kind of behavior. Now she understood the connection between the microscopic view and the macroscopic experiment she saw in the lab. The higher the temperature, the larger the motional energy of the molecules, the more violent the collisions between molecules, and the larger the intermolecular space between the molecules. She could easily imagine that if the liquid in the lab were sealed by a movable piston, the more energetic molecules would hit the piston harder causing it to move upward, and the *macroscopic* volume would increase as she had observed.

Repeating the same experiment with water over the same range of temperatures, she observed something quite different. She started at 0°C, just a little above melting point, and gradually increased the temperature by turning the T dial clockwise. In the case of alcohol, the molecules got more energetic as the temperature increased, and as a result of the more energetic collisions the average distance between the molecules increased. But the water scene was quite different. Turning the dial clockwise, she noticed that the increase in motional energy of the water molecules was much more moderate than in the case of alcohol. Instead, molecules that were initially bonded together were now torn apart, and the original *open structure* became more close-packed, effectively *decreasing* the average intermolecular distance.

The different behavior of the two liquids was now crystal clear. Alice summarized her findings as follows: In the case of methanol (as well as any normal liquid), increasing the temperature causes the molecules to be more energetic. The molecules push one another with increasing force, and therefore the average distance between the molecules increases.

In water, the starting situation, at 0°C, was an 'open structure' similar to that of ice. However, at the same time, the 'open structure' of the ice is gradually broken, and molecules now enter the cavities and the tunnels that are empty in the ice-like structure of the low-temperature water. Thus, there are two competing effects: the first tends to increase the volume (as in normal liquids) while the second tends to decrease the overall volume of the liquid. It turns out that between 0°C and 4°C, the second effect wins out. So the net effect is shrinkage of the volume with increasing temperature.

Alice was ecstatic that she could understand one of the most outstanding properties of liquid water. She remembered Professor Holmes' first lecture when he had described the macroscopic and the microscopic views. She recalled how she had been lost in that terminology. She finally felt that she had experienced both views and she no longer had any qualms about the mystery of the 'microscopic view.' That day, the two views became very natural.

Alice went back to the lab and saw Professor Holmes clearing up the chamber where the macroscopic experiment had been carried out.

"I hope that you are now comfortable with the two views of matter — the macroscopic and the microscopic," said the professor.

As usual Alice was amazed how the Professor always seemed to know what she was thinking. She wondered whether he had secretly invented another machine that could penetrate her brain and know her thoughts. But overwhelmed by the day's discoveries, she decided not to bother asking the professor how he could read her mind. "Perhaps I will ask him about that some other time," she said to herself.

As she was preparing to leave, Professor Holmes added, "You've had the benefit of experiencing both the *macroscopic* view and the *microscopic* view. You should know, however, that this luxury was not available to scientists of the 19th and early 20th centuries."

"At that time scientists understood very well the macroscopic properties of water, and in particular, the anomalous temperature dependence of the volume. But they could only guess what the microscopic view might look like. To explain the unique temperature dependence of the volume of water they postulated that water may be viewed as a mixture of two components. These are not really two different components like, say, a mixture of water and ethanol, but two components of *water*, differing in their *local* structure. One component was similar to ice — hence, having an open structure. The second component was like a normal liquid, and thus more close-packed on a molecular level. This theory was known as the *mixture model* approach to liquid water. You should realize that the characterization of the two 'components' was done before scientists knew anything about the microscopic view of matter, in general, and of water, in particular."

"The explanation of the temperature dependence of the volume is very simple in terms of the mixture model approach. As we heat the water, some of the ice-like component 'melts down' to form the more close-packed component. This transition involves a *reduction* in the volume of the overall mixture. At the same time each component *expands* with the increase in temperature. So, you see that there are two competing effects: an overall expansion of the two components and a melting of the ice-like, open structure. The first wants to increase the volume, the second to decrease the volume. It turns out that between 0°C to 4°C, the second effect is the stronger one, leading to a net decrease in the volume of water."

"In addition, the mixture model provides you an explanation why we refer to liquid water as a 'structures liquid.' In ice all the molecules are 'structured,' of course. Once the ice melts, the water becomes a mixture of two components: one structured and one non-structured. When the concentration of the first component is large, we can say that the overall liquid is more 'structured.'"

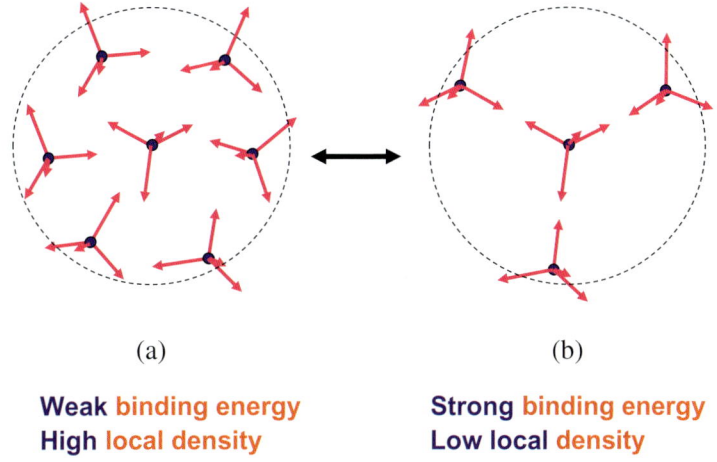

Fig. 8.3 Two forms of water having different local densities.

"This behavior is unique to liquid water. Of course, one can apply the mixture-model view to any liquid. For example, one can always *define* two components, one with a low local density, and another with a high local density, (Fig. 8.4). However, this approach will not be useful in terms of understanding the properties of normal liquid."

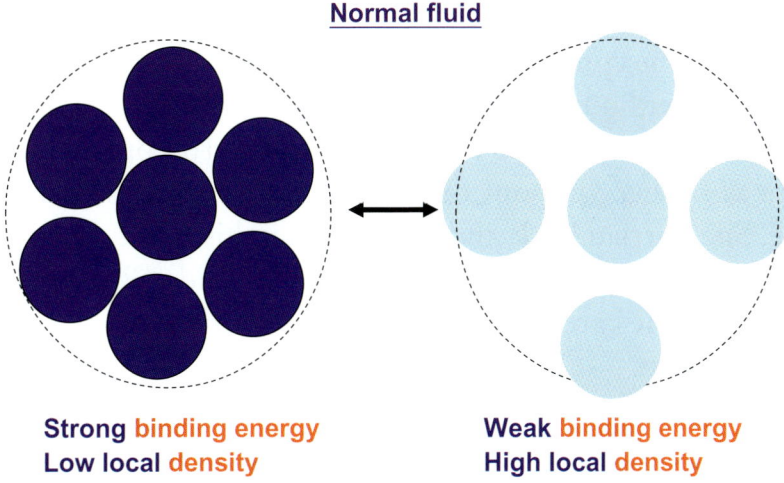

Fig. 8.4 Two possible distributions of molecules in a simple liquid.

Fig. 8.5 The effect of pressure on ice.

Alice was perplexed by what she heard. She realized how fortunate she was to have experienced the microscopic view first hand. She also realized that her understanding of the behavior of water would not have been possible without experiencing the microscopic view. "Thanks to the shrinking machine," she thought to herself. She could not imagine how scientists could understand the behavior of water without the privilege of the shrinking machine — or any similar device that allows one to experience the microscopic view.

As Alice was on her way out of the laboratory, Professor Holmes called out her name. She hurried back and found him in his office, busily buffing one of his trophies. Without looking up, the Professor said, "Remember the phase diagram of water?" He carefully returned the trophy to its place, satisfied with its newly buffed sheen.

"Suppose you start with any point on the solid phase region in the phase diagram. Then you increase the pressure until you hit the liquid phase. As you know, this can happen because the solid-liquid equilibrium line has a negative slope. Normally, this line has a positive slope, which means that applying pressure in the liquid phase can produce solid, whereas in ice the phenomenon is the reverse."

"I suggest that you do a simple experiment that will amaze your audience. Take a big slab of ice, making sure that its temperature is very low and also that the surrounding temperature is below the freezing point of water, say, $-5°C$. Take a very thin metal wire with heavy weights at both ends, and lay the wire on top of the slab allowing the two ends to dangle either side. What do you think will happen?"

Alice had no idea what the Professor was driving at, and she was totally at a loss as to what she should say. Sensing Alice's discomfort, Professor Holmes explained: "Suppose you hung that same thin wire with the weights on your finger, how do you think it would feel?"

"Of course, it would hurt! Worse still, it might even cut my finger," replied Alice.

"And why is that?" the Professor asked.

"If the weights were heavy, they would pull down the wire, exerting tremendous pressure on my finger. That would surely hurt."

"Now that you know what would happen to your finger, let's go back to the experiment with ice," suggested the professor.

"Well, if the weights are heavy enough, they will pull the wire downwards, exerting pressure on the top portion of the slab, which in turn will result in the wire's burrowing into the ice. If the pressure were high enough, it might even cut through the ice, the same thing that could happen to my finger. The only difference is the ice wouldn't get hurt," Alice added with a chuckle.

"That's more than correct!" Professor Holmes said, beaming with a sense of victory. "Indeed, the wire would dig into the ice, but the ice would not *feel* anything. In fact, it would not even split into two pieces. You are right: the experiment would not hurt the ice."

Alice wasn't sure where this conversation was going. When she said the ice would not feel pain, she meant it figuratively — only imagining that the ice had the capacity to feel. Had the Professor taken her words literally? Anyway, what did all this have to do with the phase diagram of water? Even more confused, she tried to shake off the ridiculous idea, "does ice feel?"

"Of course, Alice, the ice does not have any feelings," said the professor at last, amused at his student's perplexed expression. "Unlike your finger, the ice would never *feel* the pain and pressure exerted on its surface. What I meant when I said that the wire will not 'hurt the ice' is that it will just cut through the entire slab, but the ice will remain intact as if nothing has gone through it! In this sense, the ice will not 'feel' anything.

"This phenomenon has something to do with the phase diagram of water. As the wire presses hard on the ice, the immediate region under the wire melts. This is a result of the negative slope of the solid-liquid equilibrium line. Once a small region of the ice becomes liquid, the wire can move downwards into the ice. With the continued pressure, the layer beneath melts and allows the wire to move further downwards, and so on. If the wire is thin enough and if the ambient temperature is lower than $0°C$, then the liquid water just above the wire will refreeze immediately after the wire has passed through it. Thus, in effect the wire passes through the ice by melting it on its way down. Eventually, the wire goes all the way through the slab, yet the solid ice remains intact. It's as though no 'injury' were inflicted upon it — quite a different scenario to what would happen to your finger!"

"The same ideas might apply to ice-skating and partly explain why ice is so slippery. One explanation could be that the blades of the skates exert pressure on the ice causing it to melt locally, which makes it slippery. Still, that's only one explanation for this phenomenon; scientists are unsure about the exact explanation."

Like an exultant child given with a beautiful toy, Alice felt that she had indeed gained a treasure trove of knowledge. Now, everything she saw somehow had more meaning and significance, more than just her eyes could see — especially because she alone could experience the microscopic view from the 'inside.' She headed home, with a joyful spring in her step.

9
The Heat Capacity of Water

On the way to the laboratory, Alice saw a billboard advertisement of a new fruit juice mix. Beneath the mouth watering visual was a slogan: BEAT THE HEAT, JOIN THE HEAP. Her imagination transported her to the beach, sipping the refreshing concoction.

The loud honking of a car horn startled her, and she realized she had been daydreaming. "I must hurry," Alice thought. "Professor Holmes is attending a faculty meeting after our discussion, and I must not be late."

As she was about to cross the street, a delivery truck slowed down as it prepared to make a turn at the next corner. From the truck's markings she knew that it was a seafood distributor. Constantly curious and observant, she quickly scanned the truck and read the phrase "*Maximum load capacity: 1 ton*" printed just above the truck's rear fender.

She was becoming annoyed with herself for getting distracted — first the billboard and now this truck — when it suddenly occurred to her that the words she had just seen, 'heat' and 'capacity,' were exactly what she was going to learn about today: *heat capacity*. She knew what each of them meant separately, but together? She had not the faintest idea. Did they have anything at all to do with her last exploration? Professor Holmes would clarify everything, Alice thought, and she took big, hurried steps in the direction of the laboratory.

Although she loved being in Professor Holmes' laboratory, Alice preferred his adjoining office, where things were more organized. The lovely orchids sitting on the window sill were always a welcome and refreshing sight. When

he was not in the laboratory, she would always find him here. But the professor was nowhere in sight.

"I hope I did not miss him," she muttered, blaming herself. "If only I had not stayed so long looking at that billboard." She was more than a little disappointed at the prospect of missing the opportunity to learn something new today.

Resigned to having missed the professor, Alice decided to leave. On the way out of the laboratory, her eyes wandered again, settling on the mysterious portrait. Every time she looked at that picture, she had the feeling that she had seen that girl before. She looked so familiar. Alice always had an overwhelming urge to greet the girl in the frame. Could it be that she was only responding to the girl's greeting?

Suddenly, she heard a door shut not far from the Professor's office, and footsteps approached.

"Of course! There's a storage room just after the last lab bench. Professor Holmes must have been in there," Alice thought, relieved that she had not missed him after all.

"There you are, Alice," said the professor cheerfully, wiping off the dust that had settled on the boxes he was carrying. "The faculty meeting was moved to tomorrow, so I thought I'd check the storage room for some materials that I might need for my classes. I've already prepared some charts and diagrams for you. They're on top of the counter right before my office. Go and get them, would you?" Alice did as she was asked and the professor went on.

"Today, there is no need to do any experiments, either in the lab or in the 'interior' of water. You know enough to understand the concept of *heat capacity* of water and its relevance to the temperature control of our body."

"As you already know, heating causes an increase in the temperature. In molecular terms this simply translates into: adding thermal energy to the system causes an increase in the (kinetic) energy of the atoms and molecules. That is clear. However, different substances react differently to the additional thermal energy."

"Think about pushing a bowling ball; the heavier the ball, the more energy you need to set it in motion or to accelerate it, if it is already moving. Likewise, different molecules need different 'pushes' to be accelerated. The *heat capacity* is a quantity that measures the rate of change of the temperature upon addition of a unit of energy; alternatively, it measures how much energy is needed to raise the temperature of, say, one gram of liquid, by one degree Celsius. Clearly, in this definition, neither the word 'capacity' nor the word 'specific' are mentioned. However, the amount of heat (or thermal energy) required to raise one gram of substance by one degree Celsius varies for different substances."

"To understand why the term 'capacity' is used, consider different jars of various shapes and sizes. Clearly, the larger the jar, the larger its *capacity* to contain, say, water. However, if we asked how much water, or any other liquid for that matter, is needed to increase the *level* of the water in the jar by one centimeter, then this quantity would be *specific* to each jar." Professor Holmes paused, allowing Alice to absorb what he had just said, and then went on.

"In the same vein, people have believed through the years that heat is some kind of fluid that flows from one body to another, somehow 'occupying' the substance."

"Clearly, we can add heat as much as we want to a given substance. However, if we *specify* the substance, let us say, water; the amount, say, one gram; and the increment in temperature, say 1°C, then the amount of heat required to increase that specific amount of a specific substance by 1°C is *constant*. This amount is called the specific heat, or the heat capacity. The larger the heat capacity of a substance, the larger its *capacity* to absorb heat. Therefore,

more heat is needed to increase its temperature by one degree. You see, the concept of capacity is used here in a figurative way — as if the substance has the capacity to 'contain' the amount of heat."

"In contrast to the example of the jars that we filled with water where you could see the water occupying the volume of the jar, you cannot *see* heat that is transferred to the substance. Yet the heat is 'gobbled up' by the water, and as a result the temperature of the water increases."

"Let me remind you of one of your earlier experiments in the vapor phase. As I have explained, turning the T dial in most cases causes a change in temperature. This is expressed as a change in the average speed of the molecules. You should recognize, however, that what you actually do in the laboratory is to supply heat to the system by placing a heater or a burner under the system. When we say that we *heat* the system, we mean two different things: one is a flow of heat into the system; the second is that the system is being heated, i.e., it becomes hotter. These two things are not always the same thing."

"You can transfer heat to a system where the temperature stays constant. Let me explain with two examples."

"Consider first the pure gas of argon. When heat flows into the system, all the energy that enters into the system causes the atoms of the argon to be more *energetic*, i.e., they travel at a faster speed on average. This increase in the average speed is also expressed as an increase in the temperature. Thus, in this example, we always find that *heating makes the gas hotter*."

"Another extreme case that you have experienced is when you have ice at equilibrium with water vapor. You remember that when you were at point X in the phase diagram and you turned the T dial clockwise, there was no change in temperature? What happened was by turning the dial you were supplying *heat* to the ice. This heat was absorbed by the ice to break the bonds between the water molecules and send the molecules into the gas phase. We say that the energy supplied causes an evaporation, or *sublimation*, of the water molecules from the solid to the vapor phase. In this case, the energy that was imparted to the system did not cause an increase in the temperature. It is often said that when there are two phases at equilibrium, the heat capacity is *infinity*."

When Alice heard that heating could both cause and *not* cause a change in temperature, she wondered whether it was possible that 'heating could also cause cooling.' What the professor said next made that idea sound rather awkward.

"Heating in the sense of heat flow into the system can either increase the temperature or leave the temperature unchanged. It can never cause a *decrease* in temperature. The reason is quite simple. Think of the energy that flows into the system as being split into two parts — part one and part two. Part one goes to breaking the bonds between the molecules. Part two goes to increasing the kinetic energy of the molecules, which in turn is expressed as an increment in the temperature. Clearly, for a given amount of heat flow, the larger the amount in part one, the smaller the amount in part two will be, and hence, also the smaller the increment in the temperature. The extreme case is when all the heat goes into part one, leaving none for part two. Obviously, part one cannot be larger than the total energy that entered the system, and therefore a zero change in the temperature is the *minimum* possible change in temperature. This excludes a negative change in temperature."

Alice once again delighted at the Professor's uncanny ability to answer to a question before she managed to ask it.

"Let us summarize these two extreme cases," he continued. "In the case of pure gas, you add a certain amount of heat, which causes an increase in the kinetic energy of the molecules. This is expressed as an increase in temperature. In the pure solid at equilibrium with the vapor phase, you add the same amount of heat, all the heat energy is absorbed in breaking the bonds between the molecules. Hence, there is no change in temperature."

"In liquid water, we have an intermediate case. When we supply a fixed amount of heat to the water, some of the energy is absorbed to break the bonds, or 'melt' the ice-like patches of water molecules. The remainder is translated into the kinetic energy of the molecules, which is expressed as an increase in temperature."

"This is exactly what you observed in your last experiment and which you aptly described as the 'sluggish response of the molecules' when you turned the T knob. What you did by turning the T knob clockwise was *not* to increase the temperature directly, but to *heat* the system. If the whole heat energy was transferred into the *motional* energy of the molecules, you would have seen an immediate change in the average speed of the molecules, which is expressed as an increase in the temperature."

"However, when part of the heat energy is 'wasted' in breaking the bonds between the water molecules while the rest goes into the motional energy, clearly you will see a sluggish response and the temperature will go up slowly. That is what you *observed* and that is exactly the meaning of the heat capacity of water."

"To increase the temperature of 1 gram of water by 1 degree Celsius, you need more heat energy than if you were to increase the temperature of 1 gram of alcohol or oil or benzene by 1 degree. Why? Simply because a large part of the energy goes to the breaking of the bonds between the water molecules, and a small part of the energy goes into the motional energy of the molecules, and hence, into the increasing of the temperature."

"Because a relatively large part of the heat supplied to water goes into breaking the bonds rather than to the motional energy, we need more heat to increase the temperature of 1 gram of water by 1 degree Celsius. This is the reason why the heat capacity of water is larger than that of most liquids."

"By the way, the term *calorie* is defined as the amount of energy required to increase the temperature of 1 gram of water by 1 degree Celsius. There is the *small calorie*, or gram calorie (symbol cal), which is the amount of heat required to increase the temperature of water by 1°C (say, from 15°C to 16°C). We also have the *large calorie* or kilocalorie (kcal), which is 1,000 small calories.

Of course, this quantity depends on the temperature and pressure. The heat capacities of water at 0°C, 10°C and 25°C are different, but we shall not be concerned with such differences."

"Incidentally, the term *calorie* is also used in labeling the 'energy content' or the 'food energy' associated with a particular food. This caloric value is related to the amount of energy *contained* in one gram or one kilogram of that food."

"I should also add that the molecular, or microscopic, understanding of the heat capacity was not easy to achieve. You had the privilege of using the shrinking machine to experience the microscopic view. In the past, scientists were able to explain the large heat capacity of water by using the mixture model approach. You recall the mixture model that was invoked in the explanation of the temperature dependence of the volume? The same mixture model can be used to explain the relatively high heat capacity of water."

Alice felt fatigue slowly creeping in as she listened to Professor Holmes' long lecture. Her joints ached, her nose had started to run, and her eyes were getting watery. Much as she wanted to hear and learn more, she knew she had to go home and rest.

"I'm so sorry Professor Holmes, but I think I'm getting a little sick," Alice told the Professor apologetically. "Perhaps we can discuss today's topic further some other time?"

"I was about to discuss the very important topic of the body temperature and its regulation, but it would not be good to do that now. We can discuss it at a later date, when you feel better. I hope you did not catch a cold, or should I say, I hope you did not catch a *hot*!" the Professor replied with a chuckle."

Alice went home feeling miserable. She was sure she had a cold. It must have been the changeable weather. It was funny, she thought, although it was no laughing matter. She knew she had a cold but she didn't feel cold at all — quite the opposite, in fact. "How come I feel hot inside when I actually have a cold?" she asked herself.

When she got home, Alice took her body temperature and confirmed that it was a little higher than normal, which is about 37°C (or 98°F). She recalled the Professor's parting words — perhaps she had a *hot* rather than a cold? Although she was feeling miserable and weak, she was intrigued by what the professor had said before she left. She knew that her 'normal' body temperature was about 37°C, but how did the body maintain that temperature and what did it all have to do with the specific heat of water?

10

The Latent Heat of Water and Regulation of the Body's Temperature

"Oh mom, we don't need to go to the doctor. It's just a cold," Alice protested.

"We don't know if it's really just a cold, and we don't want to take chances," her mother said. "So you'd better get dressed as I have made an appointment with Dr. Edwards at 2:00 pm."

When Alice and her mother got to the clinic, there was already a long queue of patients waiting for their turn to be called. Most, if not all, were noisily blowing their noses while a few sneezed incessantly. To make matters worse, the man who sat beside Alice had a hacking cough.

"Oh my! Even if I weren't sick, I think I would soon get sick with all these people around me," Alice thought.

It was finally Alice's turn. After a thorough check-up, the doctor assured her that it was nothing serious and that the fever was caused by her inflamed tonsils. He told Alice that the body keeps the temperature at around 37°C (98°F) and that sometimes when it is cold a person is more susceptible to viral infection. When the body wards off the invading virus, the body temperature goes up in the process, and so the person feels hot rather than cold. With that, mother and daughter thanked the doctor and bade him farewell.

The next day, Alice went to Professor Holmes' laboratory earlier than usual.

"I didn't expect to see you today," the professor said when he saw Alice. "Are you well now?"

Rainbow over Jerusalem.

"Oh yes, Professor, I'm much better. I'm ready for more lessons!" she replied. "I am curious to know how the body maintains a fixed temperature. Does that have anything to do with the properties of water?"

"That," said the professor, clearing his throat, "is a very important question. In each of our cells, hundreds or even thousands of chemical reactions take place. Some produce heat. Colloquially, we say that the body 'burns' food, and the energy that is stored in the chemical bonds of the molecules, such as sugar, is transformed into heat. If all the heat produced in our cells were converted into motional energy of the molecules, the body temperature would increase to dangerously high levels. The body can operate normally within a limited range of temperatures around 37°C. If the temperature goes beyond 40°C, many of the molecules, such as proteins, may be destroyed. This is called protein *denaturation*, and in extreme cases it can be fatal."

"Fortunately, most of these heat-producing, or *exothermal*, reactions take place in an aqueous environment. As we have seen, water has a high *heat capacity*, which means that water is an effective 'capacitor' or 'absorber' of heat; so only a fraction of the heat produced by the chemical reactions in the cells is used to increase the temperature of the body. This is an important factor in the regulation of the body's temperature."

"There is another well-known effect that helps to regulate the body temperature. Almost everyone has experienced that when the ambient temperature gets very cold, the body shivers involuntarily. This shivering can be likened to doing exercises, the body 'burns' some high caloric substances and produces heat in the process, increasing the temperature of the body until it brings it back into the normal range. Another important mechanism used by the body to regulate its temperature is sweating. Here, it is the *heat of evaporation* that is the main protagonist."

"Now, let me first explain the concept of latent heat. This term was introduced at a time when scientists believed that heat was a kind of fluid that flowed from one body to another. In solid or ice, it was believed that this 'fluid' was *latent*, or hidden."

"Today, one speaks of the heat of vaporization, the heat of fusion and the heat of sublimation."

"When you were experiencing the two-phase region, you turned the T dial clockwise, but the temperature did not change. What actually happens is that you heat the ice or the liquid and the energy that flows into the system breaks the bonds between the molecules. When liquid is in equilibrium with vapor, all the heat that flows into the system is used to overcome the bonds between the molecules, and sends them into the gas phase. The same occurs when you heat a solid at equilibrium with gas: again, all the heat that is supplied is used to 'evaporate' — we sometimes say 'sublimate' — the molecules from the solid into the gas."

"When ice is at equilibrium with liquid water, and we supply heat, again this energy is used to break the bonds between the molecules, but unlike the process of evaporation, not all the bonds are broken. Estimates have suggested that about 15–20 percent of the total bonds in ice are broken when ice transforms into liquid."

"It is well known that the heat of vaporization of water is quite high compared to other liquids. This fact is exploited by the body in regulating its

temperature when the ambient temperature is high, or when you do intensive exercise, your body temperature tends to increase. Perspiring is one mechanism to compensate for this increase in temperature. Under the skin are *sweat glands*, which under 'hot conditions' release sweat — mainly water — to the surface of the skin. The evaporation of this water requires energy, and this energy is 'taken away' from the body, rendering it cooler."

"Due to the high value of the heat of vaporization, water is quite efficient at cooling the body. A relatively small amount of water is lost to produce a large cooling effect. In fact, when you sweat and there happens to be a breeze, you feel a mild brushing of the cold air on your skin. This is the same effect as blowing on hot soup, tea or coffee: the air that is blown onto the surface of the liquid causes evaporation, which in turn causes cooling."

"Note, however, that it is not the temperature of the air from the blower's lungs that cools the soup. That airstream blowing over the surface of the soup could be warm, but it 'grabs' some water molecules from the surface of the soup and transports it to the gas phase. This process requires energy that is taken from the bulk water in the soup or the tea — hence the cooling effect."

"It should also be mentioned that the *rate* of the sweat's evaporation from the skin depends on the *relative humidity* of the air around the body. Relative humidity is a measure of the amount of water molecules in the air relative to the maximum amount of water molecules that the air can hold at that specific temperature. The smaller the relative humidity of the air, the larger its capacity to accept more water molecules."

"When the relative humidity is 100 percent, the air cannot contain any more water molecules. When the air is very dry, water evaporates quickly, which explains why laundry dries quickly in dry weather with low relative humidity. On the other hand, when the relative humidity is high, the air's capacity to absorb water is low; thus cooling becomes less efficient. A good example would be a humid area near the sea when sweat sometimes stays on the skin much longer, which makes us feel sticky."

"Occasionally, there are patients in hospitals who are at risk of life-threatening conditions as a result of very high temperatures. To lower the patient's body temperature quickly, a large quantity of alcohol is poured onto the patient's body and a fan is used to cause evaporation. This creates the same effect as blowing on the surface of hot soup."

"Why use alcohol rather than water to cool the body? Does alcohol have a higher *heat of evaporation*?" asked Alice.

"That is a good question," replied the professor. "In fact, the heat of evaporation of water is larger than that of alcohol. The reason for using alcohol instead of water in such instances is not because it has a higher heat of vaporization, but rather because alcohol vaporizes more quickly, and thus can bring the body temperature down faster."

Fig. 10.1 The drinking bird.

Alice was feeling overwhelmed again. Although she now realized how important the properties of water were for regulating body temperature and she had understood everything that the professor had said, somehow she was finding it hard to take everything in. She needed to have a break and get some fresh air. Alice's gradual change of mood had not escaped the professor's attention, and he quickly switched the conversation to a lighter subject.

"An entertaining but sometimes puzzling toy is the 'drinking bird,'" he began. "The toy operates based on several physical and chemical principles. One of these is the cooling effect of water evaporation. I will not dwell on the details, but when you feel better try to figure out why the bird seems to be drinking forever."

This toy had long fascinated Alice, and she had seen it operating many times without ever having really understood how it worked. On her way home, she thought a great deal about the drinking bird — and whether it would go on drinking forever. She decided to examine its inner workings in more detail when she got home.

11

'The Electrified Water'

It was a glorious day at the beach. Alice squatted on a mat and read a novel while her cousin Matthew, a three-year old with big, aquamarine eyes and light yellow curly hair, was busily shoveling the sand. He would use either his toy shovel or his tiny, stubby hands to gather handfuls of sand, carefully depositing his collection inside a toy pail.

"Fly, fly, fly away!" Alice heard Matthew exclaim gleefully. The last thing she saw was the pail and the shovel flying in the air, and the upward sweeping motion of Matthew's little hands. Before she could react, the wind had swept a few grains of sand into her eyes. She felt a sudden burning sensation and ran and dunked her head into the sea in a panic.

"Yuck!" she exclaimed as she emerged from the water. She had managed to dunk her head with her mouth open and swallow a little seawater. "Oh my! That was awful! Why didn't I just wash the sand off with fresh water?" she said, frustrated with herself.

Having finally got the sand out of her eyes, she swam back to the beach only to find Matthew shoveling again as if nothing happened. But she just couldn't get mad at him. She just sat a little further away lest he try his 'flying' stunt again.

As she sat admiring the water's endless horizon, a thought struck Alice. "Why was there so much talk about the shortage of drinking water when there's so much seawater right under our noses? Why can't water from the ocean be used instead? Why do we constantly have to be reminded to save water? This is puzzling!"

When Alice arrived at the laboratory the next day, Professor Holmes seemed to have been reading her mind again.

"Today, we shall explore the electrolyte solutions," he began. "These are the solutions that you encounter at the beach. Electrolyte solutions are obtained when salt is dissolved in water. There are many kinds of salts, but for simplicity you can think of table salt — the common salt that is also known by the name sodium chloride. Here I have prepared a simple experiment. There is pure water in this glass and some crystals of salt, sodium chloride, on some paper."

The professor handed Alice a one-buttoned gadget and instructed her to tell him when she was ready to start. "I suggest that while I add the salt into the water, you watch what happens from the 'inside.'"

This was exciting! Alice was eager to find out what really happened when salt was added to water. All she could tell from the 'outside' was that the salt simply *disappeared* in the water. But where did it disappear to? She was anxious to know.

In no time at all, the shrinking machine had taken her back to the microscopic world of pure water. It was another case of déjà vu. She adjusted her heavy goggles and settled herself on one of the oxygen atoms, looking around to see how the coordination number would fluctuate. Mostly it was exactly four; sometimes, other water molecules invaded the empty space around her, temporarily increasing the coordination number.

She was ready to see what happened as salt was added. Squeezing the button on the gadget that the professor had given her, she waited and looked around her impatiently.

The water molecules had begun to move strangely, and looking around she saw two kinds of spheres, one small and one much larger. When the spheres came nearer, she noticed that the water molecules around the small spheres were arranged quite differently to the way they were arranged around the big one. Occasionally, when her molecule came close to the small spheres, she noticed that the molecule was oriented in such a way that the oxygen touched the sphere while the hydrogen atoms were pointing away. Around the bigger spheres, however, the orientation was such that the hydrogen atoms quite often pointed *towards* the sphere.

Fig. 11.1 A dipole mement and the resultant dipole moment of water.

She also noticed that the arrangement of water molecules in her surroundings was now quite different from that of pure water. In the immediate vicinity of each sphere, the water molecules seemed to be locked in an almost permanent structural arrangement. In contrast, in regions farther away from the spheres, there was almost total chaos — with no discernible structure at all.

Alice remembered Professor Holmes saying that he will add salt to the water. Did he add two salts; one small and one large? She was eager to ask the professor what the two salts were (Fig. 11.1).

As soon as she got out of the shrinking machine she heard the professor giving his explanation before she had a chance to ask.

"That was common salt — sodium chloride. When salt is added to water, the sodium and the chloride ions separate. Each molecule as a whole is electrically neutral, but when they are dissolved in water, they are no longer neutral atoms but rather charged particles. The sodium is positively charged, indicated by a plus sign in the figure. The chlorine atoms, on the other hand, are negatively charged, indicated by a minus sign. These charged atoms are called *ions*. Positive ions like the sodium (Na^+) are called *cations*, and negative ions like the chloride (Cl^-) are called *anions*. The structural rearrangement you have observed is a result of the strong interaction between the electrical charges on the ions and the electric *dipole moment* of a water molecule."

"Let me explain. A water molecule is a neutral entity as a whole. However, like any atom or molecule, the water molecule comprises nuclei that are positively charged and electrons surrounding the nuclei that are negatively charged."

"As you can imagine the distribution of charges is such that it is not spherically symmetric. Without getting into the details of the distribution of charges, scientists found that a water molecule acts as if it were an electric dipole — that means, as if it consists of only two *point-like charges*, one positive and one negative."

"In reality, the distribution is much more complicated, but many properties of water in an electric field can be explained by treating a water molecule as a simple *dipole*, i.e., two oppositely charged particles, one near the center of oxygen and one along the line bisecting the HOH angle." (Fig. 11.2)

"The fact that a neutral molecule such as sodium chloride (NaCl) dissociates into two separate ions, each surrounded by water molecules, is evidence of the strong interaction between each of the ions and the water molecules in the surroundings. It also explains why sodium chloride is very soluble in water: as you have no doubt experienced at the beach, seawater tastes very salty."

The professor paused and then proceeded to perform his usual trick — answering Alice's questions before she got a chance to ask them.

"You were probably wondering why there is a shortage of drinking water when there is an abundance of water in the sea. Well, the answer is simple. Seawater contains large amounts of different salts. To get drinking water one needs to get rid of all the salts. This process is called *desalination*, i.e., removal of the salt from the salty water. The technology for desalination is available, but unfortunately, the process is rather costly. Therefore, until people develop a more efficient and less expensive method of desalination, we will not be able to use this vast amount of water from the seas."

"However, when water evaporates from the sea, it leaves the salt behind. The water then forms the clouds, falls as rain, and then fills underground wells. This water is relatively free of salt and therefore suitable for drinking."

"As you can see in this little diagram, the hydrogen atom (H) has a little positive charge, and the oxygen atom (O) has a little negative charge. We draw an arrow from the negative to the positive charge. What we call the 'dipole moment' of water is the *resultant* arrow, which is depicted as a heavy arrow in the figure."

"Let us see what happens to the electrical dipoles of water molecules when they are put between two plates, one positively charged and the other negatively. The water molecules will still rotate vigorously. However, if the electric field produced by the charges on the two plates is very large, the positive pole of the water molecule will tend to point towards the negatively charged

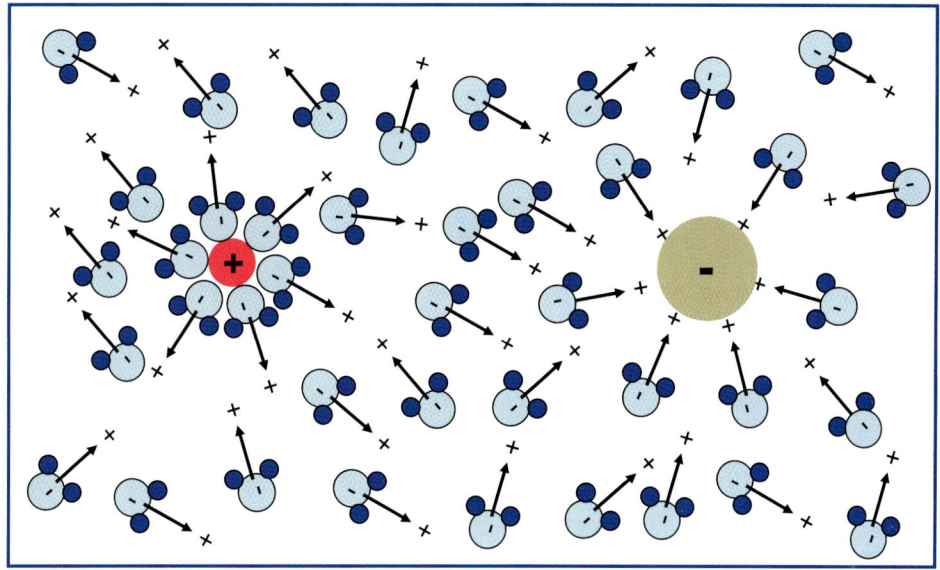

Fig. 11.2 Positive and negative ions.

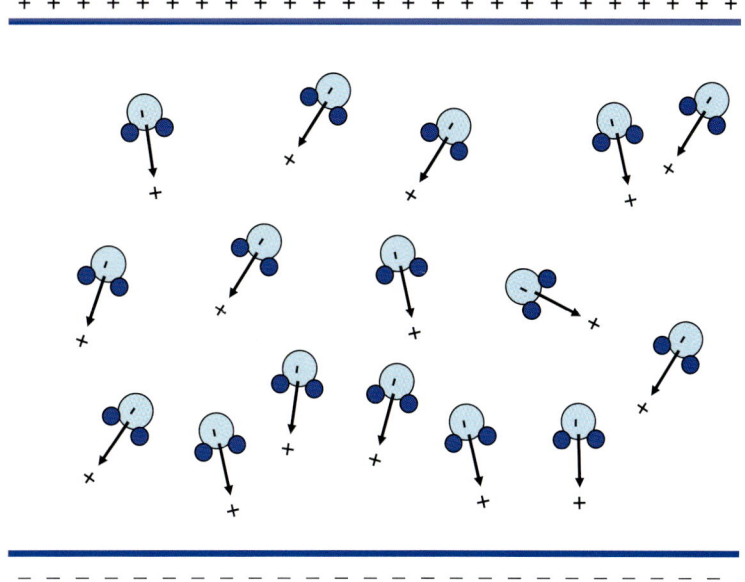

Fig. 11.3 Illustration of a water molecule in an electric field.

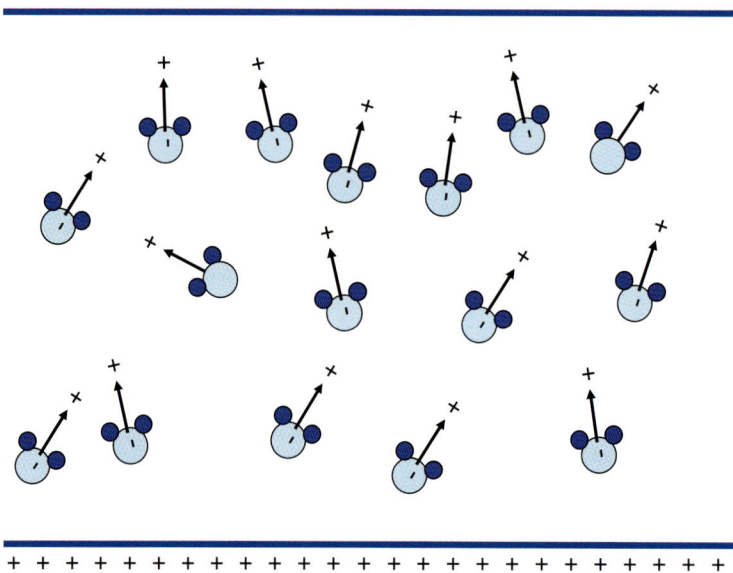

Fig. 11.4 The effect of reversing the electric field.

plate and the negative pole toward the positive plate. As you can see in this illustration, the orientations of the water molecules seem quite random, but you will notice that most of the arrows are pointing downwards — towards the negatively charged plate."

"Now, what happens when we reverse the charges on the plate? The negative charge up, the positive charge down?"

Alice had no problem with that. "Clearly, the arrows will point upwards on average — towards the negative plate."

Professor Holmes was satisfied with her answer and went on.

"The larger the electric field, the larger the preferential orientation of the water molecules. A similar phenomenon occurs around the ions in water. A positively charged ion (cation) produces a very strong electric field around it. Therefore, the dipole of the water molecule is oriented in such a way that the negative pole of the water molecule is closer to the cation than the positive pole. Since the electric field is very strong, almost all the water molecules in the

immediate vicinity of the cation are almost 'frozen' at their preferred orientation."

"At distances farther away from the center of the cation, the electric field is not so strong. Hence, the influence on the orientation of the water molecules is weaker. In this region there is a competition between the natural structure of water and the ordering effect of the electric field. The water molecules in this region get 'confused' and do not 'know' how to orient themselves. This confusion, resulting from two conflicting forces, produces almost total chaos. At still larger distances from the center of the cation, the electric field is very weak. Here, the natural structure of water dominates, and the water molecules do not feel the effect of the electric field of the cation."

"Similar effects are produced by the negative ions (the anions), but in this case the orientation of the water molecules is such that the positive pole of the water molecule comes closer to the center of the anion."

Professor Holmes paused momentarily and then asked, "Have you ever tried heating a glass of water in a microwave oven?"

"Yes, I have," Alice answered, although at the back of her mind she wondered what that had to do with ions and the electric field.

"Did you notice something strange that occurs when you heat a glass of water in a microwave — which is quite different from heating the same glass of water in a conventional oven?"

Alice was at a loss. Who heats a glass of water in an oven? How could she answer the question?

Professor Holmes continued. "Well, imagine you put a glass of water in a conventional oven and another in a microwave oven. What would the difference be?"

Alice had absolutely no idea what the professor was driving at. She had often used a microwave oven to heat food or defrost frozen food — or even to boil water to make tea. She had boiled water in the microwave on numerous occasions to fix herself a mug of hot chocolate. Once the water boiled, all she did was take it out of the microwave oven. She *had* observed that the mug was warm but never hot. But it had never occurred to her to boil water in a normal oven! She knew it would take much longer, and the glass of water would probably be too hot to handle.... It was then that it dawned on her that something *strange* happened when one heats a glass of water in the microwave. Why indeed did the water boil inside the mug while the mug itself remained relatively cold?

"Did you ever think about the strange way that water is heated in a microwave oven?" the professor asked. "The water heats quickly and yet the container stays cold, as if the water is being heated from the inside rather than from the outside."

"That is indeed puzzling," said Alice. "I never gave it a thought. I do not know how the microwave oven operates. Does it heat only the water and not the container?"

"That is exactly what happens," replied the professor, "and it has something to do with the electric dipole moment of water. What the microwave oven does is produce an electric field that changes direction many times every second. The electric dipoles of the molecules try to orient themselves in the direction of the electric field. But the direction of the field keeps changing very fast, and the water molecules have to continuously reorient themselves in different directions. Since this happens so many times per second, the rotational energy increases, and this is reflected in an increase in the temperature of the water. In

short, we can say that the oscillating electric field causes the *water* molecules to rotate faster; there is no such effect on the glass that contains the water."

"Most foods contain water. In a conventional oven, the whole interior of the oven heats up. Whatever container is used it heats up starting from the *outside*, then penetrating by heat conduction to the interior of the container. In the microwave oven, however, it is the water molecules in the food that react to the oscillating electric field. The rotation of the molecules is translated into heat. A container made of glass or paper doesn't contain water, so it isn't directly affected by the oscillations of the electric field. That explains why the container stays cold — but not forever. As the water becomes hot it also heats up the container, by heat conduction, though this takes some time. In a sense you are right in concluding that the microwave oven heats from the inside to the outside provided that what is inside contains water — or other small molecules having a dipole moment — while the outside does not. The conventional oven, on the other hand, heats from the outside towards the inside of whatever is in the container."

Alice was overwhelmed, but not only from discovering that the orientation of water molecules around ions was related to the orientation of the water molecules in the oscillating electrical field in her microwave. She was even more overwhelmed by the very fact that she could follow and understand everything that the professor had told her. On the way home, she thought about how much progress she had made since that first lecture by Professor Holmes. To think that she had almost given up.

12

Professor Holmes' Last Lecture

The semester was drawing to a close. Alice had accumulated a wealth of knowledge on water as well as its relevance to life on earth. It was the day of the professor's final lecture. Alice guessed that Professor Holmes would be summarizing the semester's lectures. When he sauntered into the classroom without his notes, Alice knew she had guessed right.

"Today I shall first summarize what we have learned on water and then we shall proceed to discuss some aspects of the role of water in biochemical systems. To understand the role of water in biology, we must learn about fundamental processes that take place in the cells of our bodies."

"We have dedicated most of the semester to learning about the properties of water and a little about aqueous solutions. We learned how to navigate in the phase diagram of water. We have not explored the whole phase diagram of water, but only the part that is relevant to life, i.e., temperatures between $0°C$ to $50°C$, say; and pressures of around one atmosphere. At higher pressures, water exhibits a rich repertoire of structures of ices — but these are of no concern to us."

"We've also learned about the most outstanding properties of water. The anomalous temperature dependence of the volume of water between $0°C$ and $4°C$, the large heat capacity, the structure of water, and also the orientation of water molecules in an electric field. All these properties are not only interesting in themselves, but are also relevant to processes occurring in every cell of our body."

"Next semester we shall learn about some important biochemical processes that take place in our bodies. We shall start with the Central Dogma of molecular biology, a topic I mentioned in my first lecture. The Central Dogma consists of a few processes, including the reproduction of the information 'written' on the DNA, transcribing this information into RNA, and then the translation from the RNA into proteins. Once the proteins are produced on the machinery of the ribosomes (which are themselves aggregates of proteins and RNA), they must fold into a specific structure in order to be able to do a specific job. Examples of such jobs are catalyzing the processes of replication of the DNA as well as the translation of the information contained in the DNA into proteins."

"In this process, as well as in many others, water is not a mere medium in which biochemical processes take place, but its participation in the processes is actually essential." Today, we know that processes such as spontaneous protein folding, protein–protein association and protein-DNA binding occur in aqueous media. Without the active role of water, none of these processes would occur. It is therefore clear that water molecules play an essential role in all of these processes, as well as in determining the specific structure of the final product.

"I am confident that what you have learned about the properties of water during the past semester will help you understand the role of water in biochemical process. I also hope this short introduction on the kind of biochemical processes that are affected by the presence of water will trigger your interest in the subject we will tackle next semester."

With these words, Professor Holmes dismissed the class and left, after wishing them a happy and restful vacation.

Alice felt very excited about the prospect of learning more about the role of water in biology. She was so grateful for having experienced the microscopic point of view, which had made the macroscopic properties of water so comprehensible. She wondered whether she would be able to use all of the professor's 'facilities' to plunge into her own cells and 'see' the microscopic picture of the processes going on. She was confident that she could make full use of what she had experienced in understanding the processes in her own body. It was time to pay the professor a final visit — at least for this semester.

13

The Last Visit to Professor Holmes' Lab

Alice's body clock had roused her from sleep even before her trusty cuckoo clock could do its job. Today she would visit the professor in his lab to clear some cobwebs in her mind regarding the various machines, as well as to thank him for everything that he had done for her. She was sure she would see him at the university when the next semester began, but she wanted to express her gratitude for all his help. Through his mentorship, she had been able to explore the fascinating world of molecules — and how they interact to produce such a unique liquid so vital to life.

But there were nagging questions at the back of Alice's mind. They had bothered her ever since her first excursions to the inner world of water in the gaseous, liquid and solid phases.

She never really quite understood how the professor always seemed to know what she was about to ask him — despite the fact that he had never used the shrinking machine and had never visited the molecular world. She had been too preoccupied with her explorations to find the opportunity to ask him how exactly the shrinking machine and IQ machine worked. How was it possible that her body — which was more than 50 percent water and contained an unimaginable number of water molecules — could be shrunk to the size of a single molecule? Each water molecule in her body would have to be reduced to perhaps a hundred millionth of its real size!

Today was particularly special because it was the last day of the semester. Alice was surprised when Professor Holmes offered her a slice of her favorite cake. It looked just as good as her mother's home-baked goodies.

"Could the professor have guessed that chocolate cake is my favorite?" Alice thought. "He seems to know everything!"

"Today I am going to divulge some well-kept secrets," said the professor, and Alice braced herself for another demonstration of his mind-reading prowess. "You have probably wondered how the shrinking machine works and how I knew about everything that you experienced, even though I have never visited the molecular world myself?"

The professor's ability to read her mind had never ceased to amaze Alice, but this time what he said shocked her.

"Indeed, I have never visited the molecular world, but neither have you!"

"What do you mean?" said Alice indignantly. "I certainly *did* visit the molecular world, as you very well know!"

"I will answer all your questions in a moment. I also hope that you will understand how I seemed to 'know' in advance everything that you experienced. As for the two machines, I must confess that I have misled you. The shrinking machine does not shrink you, and the IQ machine does not measure your IQ, nor does it provide any brain power!"

Alice could hardly believe what she was hearing. She wasn't sure whether the professor was joking or whether she never really understood.

The professor continued.

"First, as I said before, you never actually visited the molecular level. No one has ever visited that world and I very much doubt whether anyone ever will."

"For many years scientists have been investigating matter only from the outside, the macroscopic world. But they were eventually able to create a *model* of the molecular world — how that world would appear *if* we could reduce our size to that of a single molecule. These pictures were mere models, i.e., postulates, guesses and conjectures. However, as more experiments were carried out and more highly sophisticated methods of computation were developed, the molecules that were initially figments of the imagination became increasingly more realistic. All the deductions from the models were consistent with the behavior of macroscopic matter, in general, and for liquid water, in particular."

"Thus, all the things that you 'saw' were the models constructed by scientists, and those models were programmed into the machine. What you have seen is only a *simulated view* of what the molecular world would look like *if* you were reduced to a single molecule. What you have seen is just a movie. I have created a virtual world and some effects that gave you the feeling that you were actually in the movie. For instance, when you turned the T dial, I could see that on the screen of my computer, and I changed the speed of the molecules you were seeing accordingly."

"That explains how the shrinking machine works and how I knew everything that you experienced! My machine is just a simulation encapsulating the models that scientists have developed over many decades. What you saw is as close to reality as we can imagine."

As Professor Holmes' words began to sink in, Alice first felt a little betrayed. How could the professor have deceived her like that? Then she suddenly realized that it was all for her own benefit. But there was more.

"As for the IQ machine, that was also a trick!" grinned the professor. "That machine does nothing! Why did I use it, you ask? Well, from many years of experience dealing with students, I realize that there are those who have a very low self-esteem and whose minds are preconditioned. They believe that they cannot understand complicated things, and they never give it a try. They close themselves like clams and give up trying to understand anything that sounds unfamiliar, complicated or difficult. I guess you had that kind of feeling too, after listening to my first lecture."

Alice smiled sheepishly. In this instance, the professor really *had* read her mind.

"To avoid you being discouraged from trying to analyze and understand what you saw, the IQ machine made you believe that your intelligence had been given a boost. 'Conditioning' your mind allowed you to be more confident in your abilities and not be discouraged to take on difficult problems and challenges. I felt it was necessary to boost your *confidence*, not your IQ."

Alice listened in astonishment as Professor Holmes laid out his elaborate deception, but she just could not bring herself to harbor any resentment towards him. In fact, she realized just how much she admired his creativity and how he had so skillfully manage to pull the wool over her eyes. She also appreciated his honesty in telling her the real purpose of the machines.

There was a moment of silence as the professor finished speaking. He took a deep breath and gazed affectionately at the picture on the wall, the photograph that had intrigued Alice for so long. Suddenly, Alice realized why the girl had always seemed so familiar. She had the professor's eyes…

"That is Juliet, my daughter," said the professor at last. "Regrettably, she is no longer with us. She passed away as a result of dehydration on a summer camp several years ago. She loved plants. All of the plants that you see around the lab belonged to her. She lovingly took care of them every single day. After her passing, my wife and I brought them here so I could take care of them myself. That is the reason why I have dedicated many years of my life to the study of this extraordinary yet vital liquid — water."

The professor heaved a gentle sigh and turned to Alice, his sadness suddenly lifting, and he told her about his enthusiasm for what he loved most — imparting his knowledge to his students.

"I hope that this semester's lectures have somehow initiated your interest in the world of water and that you will use this knowledge in understanding the role of water in biology."

Alice nodded. "Yes, Professor. Thank you."

"So I shall say *arrivederci* like the Italians, or perhaps *au revoir* as the Parisians do, or just a simple goodbye," said the professor with a smile. "One thing is certain. We shall meet again next semester. You have been a good student Alice. I hope you will have the same enthusiasm next semester. So, goodbye for now and see you soon!"

As she walked home, Alice smiled to herself. She wished she had one of Professor Holmes' gadgets to speed up time, so she could soon be enjoying his lectures once again.